The morphogenesis of the behaviour of the domestic cat,

with a special emphasis on the development of prey-catching

J.M. BAERENDS-VAN ROON en

G.P. BAERENDS

VERHANDELINGEN DER KONINKLIJKE NEDERLANDSE AKADEMIE
VAN WETENSCHAPPEN, AFD. NATUURKUNDE, TWEEDE REEKS, DEEL 72

NORTH-HOLLAND PUBLISHING COMPANY
AMSTERDAM, OXFORD, NEW YORK, 1979

ISBN 0 7204 8480 4

authors address:

Zoological Laboratory of the University of Groningen,
Haren, the Netherlands

AANGEBODEN IN DE VERGADERING
VAN 30 SEPTEMBER 1978

Contents

I. Introduction

I.1. AIM OF THE STUDY

During many years we have watched kittens of domestic cats growing up in and around our house. The gradually changing interactions of the kittens, with their siblings, with the constantly caring mother, and with other elements of the environment intrigued us. What would be the processes by which these originally altricial kittens ultimately develop the behaviour patterns enabling them to maintain themselves as adult animals, i.e. to obtain food, to communicate with conspecifics, to defend themselves against enemies, and to raise young?

An "ethogram" of the domestic cat has been given by Leyhausen (1956[a,b], 1973, 1978). This author not only gives careful descriptions of a great many adult behaviour patterns, he also compared corresponding (homologous) behaviour patterns between different species of the Felid family. The core of the behaviour repertoire of each species consists of activities which are easily recognizable in all members of that species. Although between conspecifics differences in form of the same pattern do exist, this variation is in general smaller than that between homologous patterns of different cat species. This species-specificity must be based in the genes. This statement uncloses the problem of how the expression of the genes is actually brought about by the processes of behavioural ontogeny. Such processes must be partly similar to those involved in the ontogeny of anatomical structures; the development of behaviour, however, has an additional possibility at its disposal, i.e. the incorporation of experience.

It has often erroneously been thought that the development of species-specific behaviour would either proceed on the basis of genes without the incorporation of experience, or if experience is involved, would result from imitation (i.e. learning from examples). Berry (1908) – who by our knowledge was the first who undertook an experimental study of mouse-catching in cats – did so with this alternative as a working hypothesis. The social isolation or Kaspar Hauser experiment was actually developed by ethologists for making decisions in favour of one of these alternatives (Weiss, 1952). Their principal interest was to

---

* Since Leyhausen (1956[a]) appeared two new editions (1965, 1973) have been published. In this paper, for reasons of priority, we have always referred to the oldest edition in which the item considered was found.

demonstrate by such deprivation experiments a role of genes in the formation of adaptive complex behaviour patterns and to defy the extreme environmentalism of many behaviorists (cf. Lorenz, 1961). Apart from the work on the ontogeny of song patterns of passerine birds (see Nottebohm, 1970) the deprivation experiments have produced little evidence for cases of detailed meticulous imitation. Several of such studies have shown the importance of genes for the development of complex patterns, but they have also revealed that genes can be effective through the use of preprogrammed learning processes. Experience acquired in this way at one developmental stage in a particular context may be a building stone for another learning process at a later stage in another context. As repeatedly advocated by, in particular Schneirla (1959) and Lehrman (1953), ethologists should not limit themselves to establishing effects of genes on complex behaviour but in addition investigate how this effect is exerted; thus how the information encoded in the genes is carried to expression in behaviour patterns.

For a number of reasons a considerable influence of learning in the development of species-specific behaviour of cats was to be expected. First, there is the relatively slow development and the long period of dependence on the mother, as compared with precocial animals, such as the young of ungulates. Second, there is the evidence from several naturalists (e.g. Adamson, 1960) who, after having raised orphan carnivores with a bottle, found them to have great difficulties in catching their own prey and in general to cope with the natural environment. Third, there is the remarkable occurrence of the type of behaviour which is generally called "play" and which seems particularly developed in animals who, like the carnivores, have to develop special skills that must be performed under variable and often difficult conditions. Among these skills prey-catching is a very important one.

In this study we have set ourselves the task to describe the development of prey-catching and to experimentally test possibilities for the occurrence of learning processes in the development of the predatorial behaviour of the domestic cat. The experimental part is restricted to behaviour towards prey (Chapter III), but because of the great resemblance of prey-catching with several activities subserving other functions, such as exploratory and agonistic behaviour, a more general description of activities of cats directed against objects outside their own body – including social behaviour – will be given and problems of their development will be raised (Chapter II). The possible functions of "play" in the development of the behaviour repertoire and its adaptive application will be discussed in Chapter IV.

Already in the beginning of this century the nature-nurture issue led to studies of the problem of how in the individual kitten the ability to recognize and catch prey develops. Berry (1908) presented mice to a litter of kittens that were over 5 months old. As these cats showed an interest in mice only after the mice started to run, but dit not attempt to catch them as long as the mice were quietly moving near or even on the cats, Berry concluded that the instinct of cats only induces them to go after rapidly moving objects, but that the actual catching and killing of mice has to be learned by imitation. Yerkes and Bloomfield (1910), however, held it for possible that an instinctive tendency to kill might have waned in the cats used by Berry, because of lack of experience with such prey. This is why they, during the first 3 months and about twice a week, offered a mouse to each of 8 kittens of 2 litters, from the third week on. In contrast to Berry they found a sudden appearance of interest in mice in some kittens already before the end of the first month when they were still barely able to eat a young mouse, and in all others during the second month. One of the kittens of the first litter caught a mouse when 4 weeks old and killed and ate its first one at the age of 2 months. To the kittens of the second litter only baby-mice were given. Only one kitten, from the 4th week on, killed and ate the baby-mice. The kittens who did not catch or kill mice all showed an interest in them during the earlier presentations. However, when they were not successful from the beginning their interest seemed to wane in the course of time, and this seemed to be the main reason for the marked individual difference in success. The authors concluded that an instinct to catch and kill mice is present when the kittens are 4 – 8 weeks old, that mice are recognized at least on movement and odour and that this instinct lends itself as a basis for the acquisition of habits of dealing with mice. Although neither imitation nor experience with mice is necessary for an efficient execution of catching and killing, such conditions if present do contribute to it.

Although these experiments made it clear that kittens of one month old can manage to kill a mouse with the appropriate behaviour patterns, the results do not explain why only some of the kittens did so whereas the majority only succeeded later or failed entirely during the experimental period. This is why Kuo (1930, 1938) undertook several series of systematically designed experiments to study how the killing behaviour of kittens is influenced by different kinds of experience during the first four months of life. He investigated influences of isolation, of the presence of a rat killing mother, of the companionship of rats, of the food (vegetarian or non-vegetarian) earlier received, of hunger, of fear, of training and of different strains of rat. His results, which will be discussed in chapter IV, are mainly expressed in terms of the percentage of subjects becoming rat-killers under the conditions concerned. Because details on the behaviour of the kittens during the tests are not given, the results do not give much insight into how the different environmental factors affected the course of development of the behaviour in the individual kittens.

Such data were taken into account by Schneirla and his collaborators (Rosenblatt, Turkewitz and Schneirla, 1959, 1969; Rosenblatt and Schneirla, 1962; Schneirla, Rosenblatt and Tobach, 1963). However, they concentrated their study on learning processes during the first 6 weeks, observing the kittens in relatively small cages and before prey-catching and social behaviour were fully developed. Leyhausen (1956[a,b], 1965) has described some ontogenetical stages in prey-catching of the housecat, but he did not aim at a causal analysis of the ontogenetical development. Moreover this author did not carry out systematic experiments with freely moving prey. More recently – and simultaneously with our study – work on the ontogeny of cat behaviour was carried out by West (1972), Egan (1972, 1976) and Barrett and Bateson (1978). It will be discussed in Chapter IV.

Observations of prey-catching behaviour and its development in Felicidae in the wild are rare; the studies by Schaller on the tiger (Schaller, 1967) and the lion (Schaller, 1972), by Eaton (1970, 1972[a,b]) on the cheetah and the lion, by Kruuk (1972[a,b]) on the spotted hyena and by Elliot, McTaggart Cowan and Holling (1977) on the lion, should be mentioned here.

I.2. MATERIAL AND GENERAL TECHNICAL METHODS

Maintenance conditions. Two different situations have been used for keeping the cats:

a. At home. Only one litter at a time was kept at home, always with the mother. The cats were allowed to walk freely in our house in the countryside; the nestbox was placed in a room adjoining the garden. It was always possible during the day and mostly also during the night for the cats to leave the house and to wander about in an open area of meadows and woods, scattered with a few houses and farms. The nearest blocks of buildings were at a distance of 2 km, but closely inhabited areas in most directions were several miles away. Most likely the area covered by our cats was smaller. Casual evidence indicates that our female cats did not go further from the house than about 1 km, but that tomcats covered a larger area. The pattern of distribution is undoubtedly influenced by the presence of home-ranges of other cats, by females in oestrus etc. However, we have not systematically investigated the way they made use of the area; we only want to point out that there were only few external restrictions on their whereabouts. Twice a day food was given in the house. Moreover, the cats caught live prey (mice, voles, rats, young rabbits and birds) and were often seen to kill and handle, but not eat, moles, shrews, toads and frogs, and once a weasel.

b. In the laboratory. The cats were kept in 6 cubical wire cages (1 m a side) placed three in a row, two cages high. Metal plates in between the cages prevented visual and tactile contact between the cats. A cubical nestbox (30 cm on a side), food, water, and a box with shavings was placed on the bottom of each cage in which a triangular shelf of half the floor surface was fitted halfway on two walls.

The three bottom cages could each be connected with an outside pen, made of wire netting and measuring 2 × 1 × 2 m. The bottoms of the pens were covered with sand. Each pen contained two forked heavy tree branches and two horizontal perches on each of which a 30 × 30 cm shelf was fixed. A hanging door which could be bolted was placed in the partition from the inner cage. Visual, tactile and vocal contact through the meshes was possible between the pens. Mothers with litters were, usually from birth but always from the 4th week on, kept in a cage connected with a pen. The mother could pass through the hanging door. No box with shavings was placed in the cage; she defecated and urinated in the sand of the outdoor pen. As kittens under 8 weeks had difficulties in moving over the wire cagefloor, it was covered with newspaper during this period.

Cats without young were kept either two to four together in a cage-pen combination or alone in the top cages; the sexes were or were not separated. For a short period a larger cage with outdoor pen was used to pool eight nearly adult cats and tomcats.

Food. Cats kept at home obtained commercial types of tinned or dried catfood and milk. Cats in the laboratory were given a specially composed amorph food mixture (prescription from the animal food dealer Hope Farms B.V.; Woerden, Holland) diluted with water and kneaded to lumps.

Health. To control their condition the kittens were weighed regularly. At 12 weeks they were given the routine vaccination against cat's disease and twice a year a cure against intestinal worms. Each time after this cure the cages were cleaned with halamid and sprayed with derris. Moreover, derrispowder was rubbed into the fur of the cats. Cleaning and desinfection was postponed when females were gravid or had kittens younger than 12 weeks. During the 5 years over which the work was extended, we only once had an epedemic infection, which killed 80% of the young present at that moment.

Origin of the stock. Two generations of adult female cats were used. The first generation (4 females) consisted exclusively of cats which had grown up and lived under farm house conditions. The second generation cats (2 males and 6 females) had spent known periods at home and/or in laboratory conditions.

All cats used were very tame. They did not avoid quietly sitting or moving people and they were easy to handle. Their tameness made it possible to observe in detail without disturbance the birth of and the caring for young.

Prey-catching experiments. In our experiments on prey-catching we have used two strongly different kinds of prey to make it possible to test the behaviour of cats experienced with only one kind in first confrontations with another novel prey-species. Mice and fish were chosen because they appeared to behave without being disturbed by the presence of kittens as long as they were not being directly attacking. Apart from the fact that the prey animal had no possibility to escape in the experimental situation the encounter between predator and prey in the test room did not essentially differ from that in the field situation. A few pilot experiments with 3-week old chickens as prey had made it clear that, because of the erratic behaviour these birds displayed in the test area, standardized encounters could not be achieved. In contrast the mice and the fish moved around apparently without taking notice of the presence of kittens unless they were rapidly approaching. As the mice were surplus animals from genetical experiments and the fish mainly Cichlid fish which had been reared for ethological purposes, both had never been attacked by predators.

In far the majority of the tests the mice were either quickly caught and killed by the kittens or left undamaged (even when they had been tossed or carried around in the mouth) at the end of the test. If 10 minutes after the start of a test the kitten had not managed to capture a mouse the test was in principle discontinued; if it had captured the mouse the test lasted as long as the kitten remained interested in its prey.

Methods of observation and data retrieval. Written records were made of all observations. During the entire development of the kittens a considerable amount of 16 mm ciné-film was taken. This material was analysed with the help of an analyzing projector (Analector – Oude Delft) which made it possible to project the film at speeds varying from 1 to 25 frames per second. For drawing, single frames were projected on a horizontal surface.

The quantitative records concerning prey-catching, details of which will be given in Chapter III, were transferred to punch cards and analyzed with equipment of the computer centre of the University of Groningen.

1.3. ACKNOWLEDGEMENTS

The present work was stimulated by the descriptive and experimental analysis of the social behaviour of the jungle fowl, published by Kruijt in 1964. It was felt desirable to investigate along similar lines the morphogenesis of the behaviour of a mammal and in particular of a mammal notorious for the occurrence of play, such as the predatorial cat. The experimental work was carried out by the first author in Professor Kruijt's laboratory from 1966 to 1971; the second author only became involved in the phase of working out the results and writing down the paper. Both authors wish to gratefully acknowledge Professor Kruijt's active interest, assistance, advice and critical comments in all phases of the study.

The work would have been impossible without the constant devoted care of the animals by the late Mr. F. Bahlmann and his staff, in particular Mr. J. Holtman, Mr. A.W. Meinema, Mr. A. Nolle, and Mr. A. de Haan. Thanks are also due to Mr. J.W. Koenes and his staff of the carpenter shop of the laboratory for technical assistance concerning the experimental arrangements.

Data processing was considerably promoted by the provisions of the computer centre of the University of Groningen; we are indebted to Dr. D.W. Smits and his staff for their kind assistance in using these facilities.

For help in the preparation of the manuscript we would like to acknowledge our debt to Miss M.K. Carlstead for considerably limiting our offenses against the English language, to Mrs. H. Lochorn-Hulsebos for her unabated willingness to type and retype various versions of the manuscript, to Mr. D. Visser for the pains he took in making the best possible prints of the photographs we had taken, and to Dr. J. Veen for improving the artistic aspects of our drawings.

II. Description of the development of behaviour patterns

II.1. THE COURSE OF EVENTS IN THE PERIOD OF MATERNAL BEHAVIOUR

A couple of days before a cat gives birth the nipples swell, she looses hair and shows a marked increase of searching behaviour in which she especially is interested in dark sheltered places, preferably with a modifiable substratum. A partly covered cardboard box with newspapers on the bottom proved to be very suitable as a nesting place and was consequently always used for this purpose during this study. Nestbuilding behaviour then consisted of scratching the newspaper with the claws of the forelegs and sometimes with additional help of the mouth, thus tearing it to pieces, and of rubbing against the bottom and the walls while the cat was laying on her side. During these movements the hair lost became attached to the paper scraps and thus a lining of the box was formed. Periods of nestbuilding lasting between 10 and 30 minutes, were alternated by periods of sleeping in the box or periods of exploratory, feeding, or other maintenance activities.

Our observations of births are in full agreement with the descriptions given by Cooper (1944) and by Schneirla, Rosenblatt and Tobach (1963). Convulsions having started, the cat laid down in the box on her side, with rounded spine, the back pressed against the wall, and from time to time pushing the opposite wall through alternating stretching and bending of each of the forelegs ("treading and scraping"). With interruptions a purring sound was produced. At intervals she rose on one of her forelegs, lifting the other foreleg, to lick the genital area. Gradually the opening of the vulva enlarged, part of the foetus becoming visible. Stronger convulsions could be accompanied by deep sighs and occasionally by a loud lau-auw sound, uttered with open mouth. Each time a young slid out the mother transferred her licking from the vulva to the newborn; then the membranes broke and the young became free after which the attached placenta was discharged. Almost immediately the mother ate the membranes and placenta and continued eating the umbilical cord. Usually she stopped upto an inch from the kitten. However, in 2 of the 40 births observed under laboratory conditions, involving different mothers, the cat went on licking and eating, damaged and eventually tore the belly of the young and finally ate it. In contrast one particular cat *(br-21)* who always devoured her young started each time with the heads of the kittens (see p. 83).

The normal interval between successive births was circa 15 minutes. In such cases new convulsions started soon and, after only a brief occupation with the newly born kitten, the cat again concentrated licking on her own vulva area. When the intervals were longer (a few lasted up to 12 hours) the mother paid more attention to the young, treating it in the way described below. When convulsions started anew the young kitten had usually got hold of a nipple and hung onto it while the mother was giving birth again (Schneirla, Rosenblatt, Tobach, 1963).

When the mother had the time to direct her attention to the newborn young she licked it for longer periods, interrupted by treading and scraping. Licking affects the movements of the kitten in a way described below (p. 16).
Treading and scraping have the effect that the young are moved towards the body of the mother. The intensity of these activities varies between individual cats.

During the first two days the free living cats tended to stay with the young continuously, without eating, drinking, urinating or defecating. Only cats kept in the restricted laboratory cages might, when disturbed by the daily routine of the caretakers, leave the boxes to feed and to defecate in the outdoor pen. After two days the free living cats also went out for these purposes. Mice were probably caught on these trips, but during the first three weeks only rarely transported to the nest. Thereafter mice were daily brought in and presented to the young (see III.2.1). In the course of time the periods of absence from the nest gradually increased from ca. 15 min. a few days after birth to an hour in the 4th week, just before the kittens began to leave the nest.

Rosenblatt and Schneirla (1962) made systematic observations on the frequency of the periods of absence under the conditions they kept their cats in the laboratory (cubical cages of 3 feet on a side, where the mother could only retire to a shelf halfway up in the cage while the young had to stay on the bottom). They found absence to be very low up to the 4th week and to increase steeply thereafter.

Coming back from her trips the mother purred, nosed at one of the kittens and began to lick it, beginning at the head, and continuing on either the back or each of the sides, after which she turned the kittens belly upwards, with her nose and one frontpaw, to lick the belly and to remove urine and feces (Pl. III[b]). When the mother was finished with the kitten it struggled to get on its feet again while the attention of the mother switched towards another one. Having treated all or most of them in this way the mother laid down on her side, curved in a U-shape, her belly slightly turned upward, while the young were searching for contact with a nipple, assisted by directed licking, pushing and treading and scraping movements of the mother (Pl. I,III[d]). When satiated the young fell asleep and loosened contact with the nipple. Then the mother might stay or often also might react with rising, stretching, and leaving the nest for one of her trips. Sometimes, when the mother was showing signs of feeling ill at ease near the nestbox, she searched for a new nest site, finally transferring the kittens while holding them by the nape in the way described by Leyhausen (1956[a], p.105). After 4 weeks the kittens left the nest and gradually explored an increasing range around it.

In the beginning the kittens went not far from the nestbox although they might sleep in other places, sometimes with littermates, sometimes alone, curved in small refuges. Most of the time the mother stayed in the neighbourhood. When the young roamed about she frequently uttered a bout of consecutive *mrauw* calls, repeated with 10 seconds interval.

Our cats nursed their young regularly until the kittens were 8 weeks old. In the laboratory the kittens were not kept with the mothers for longer periods, but at home where we often kept one young with the mother for over a year we have observed suckling at irregular times between 8 and 16 weeks. A number of times we saw one year old tomcats suckling again at the time their mother was about one week before giving birth to a new litter (after having given birth the mother behaves aggressively towards earlier young toms). Schneirla, Rosenblatt and Tobach (1963) found that in mothers having a litter in the cubical cages mentioned above, the increasing tendency to leave the litter after the 4th week went hand in hand with a decreasing tendency to accept an encroaching kitten. We frequently observed as late as 8 weeks free living mothercats attempting to catch a young and persuade it to suckle by holding it under one of her front legs and licking or even biting it. Usually when the mother was feeding a kitten its littermates came and joined suckling, after which the whole group fell asleep and stayed quiet for two or more hours.

From the 4th week on the free living mother brought at least once a day, but probably on the average more often, a mouse to the nest. In the beginning she killed and ate the mouse herself. Gradually however, the kittens joined the eating (Pl. VIII), and in the 5th week the kittens handled the living mouse and might also kill it. When about seven weeks old the kittens approached the calling mother and quickly took over the mouse (cf. Leyhausen, 1956[a]). The behaviour of the kittens towards these prey will be dealt with in detail in Chapter III.

Cats living around the house had at the most two litters per year; one in the early spring, the second at the end of the summer. Only twice a mother we kept with one or more spring kittens, gave birth for a second time in the same year; in all other cases a new period of oestrus did not interfere for at least 10 months. Female kittens kept with the mother had by that time become mature themselves and several of them gave birth during the same period. Usually there was very little animosity between both cats, they might even take part in caring for the other's litter, particularly when the mother was away (cf. Ewer, 1961). Also daughters which had failed giving birth themselves were allowed to do this; in one case while a mother was giving birth the daughter was seen to eat the afterbirth (Wolda, pers.comm.). Towards sons which need longer to mature (about 15 months) the mothers acted aggressively when they came into heat. This aggressiveness subsided during pregnancy, especially in the last week before birth. After birth, however, they were no longer tolerated, and chased away from the nest box. Often such tomcats gradually disappeared from the neighbourhood. In one case a tomcat came back after three years; he proved to be still acquainted with the geography of the house. Towards humans and towards animals of other species (dogs) with which they were familiar the cats were friendly, even when the nestbox was approached.

Unfamiliar cats, males as well as females, were not tolerated in the area covered by the mother and her offspring. A familiar cat in heat, and probably with the odour of tomcats in her fur, was chased away – but two weeks later accepted. Intruders are quickly chased away, and only in rare instances when they make a stand, severe fighting may take place.

II.2. THE EARLY DEVELOPMENT OF MOTOR BEHAVIOUR BEFORE LEAVING THE NEST

Laying on their bellies young kittens are able to crawl on a substrate, which normally is the lining of the nest or the body of the mother or littermates. Basically this crawling is brought about by curving the body through contractions of the body musculature on one side progressively from head to tail. When the kitten is laid on a flat surface a curving to one side is usually after a few seconds followed by a curve to the contralateral side, the head leading this change of direction (Fig. 1[a]). A sigmoid swaying movement results (Gray, 1951). This alternating swaying movement can be continued for an appreciable length of time. In a newborn kitten the legs are able to give some support to the body and thus facilitate its movement. However, a day after the young have been nursing from the mother their body is so swollen and heavy that the legs can no longer push it up. When the kitten crawls on a flat surface the legs rest on it and tend to lag behind when the body is curved. This causes them to be sometimes bended and sometimes stretched; apparently when certain limits are passed they adjust their position by stretching or bending, respectively (Fig. 1: frames 24-29 and 60-128). These adjusting leg movements take place more rapidly (fractions of seconds) than the alternate swaying movements of head and body; they always seem to be performed in reaction to the particular external situation. However, if a newborn kitten is supported by water the leg

---

Fig. 1. Locomotion of young kittens (motion picture frames)

a) One week old; two separate sequences showing the sigmoid swaying movement. In frame 1 of the upper sequence the left forearm lays over the hand (not visible) which is bent sidewards in the wrist. In frame 24, while the head is turning to the left, this forearm is lifted and moved forwards. As a consequence in frame 29 the foot can be placed in line with the forearm. It keeps this position until frame 128 while, since frame 32, the head is moving towards the right. In the lower sequence in frame 90 the head begins to move to the left. In frame 122 the upperarm of the right frontleg is following this movement, while its forearm is lagging behind until in frame 128 the entire leg is in forward direction placed on the substrate.

b) Two weeks old. The kitten is standing on the substrate and turns towards the left. The head is leading the movement; the right shoulder and right frontleg advance while the left foreleg is still kept in position (frame 100).

c) Three weeks old. The kitten is walking on the substrate. In the frames 1 to 50 the head is moving towards the right, while the hindleg is lifted, moved forwards and put down. Thereupon, at about frame 60 the head starts its move to the left; the body follows this movement as a result of which the weight is brought to the left side, while the right frontleg is placed in a new forward position (frame 98). Frame 120 represents the mirror image of the movement described above.

N.B. The film speed was 24 frames/second. The dashes in each frame indicate the same spot on the substrate.

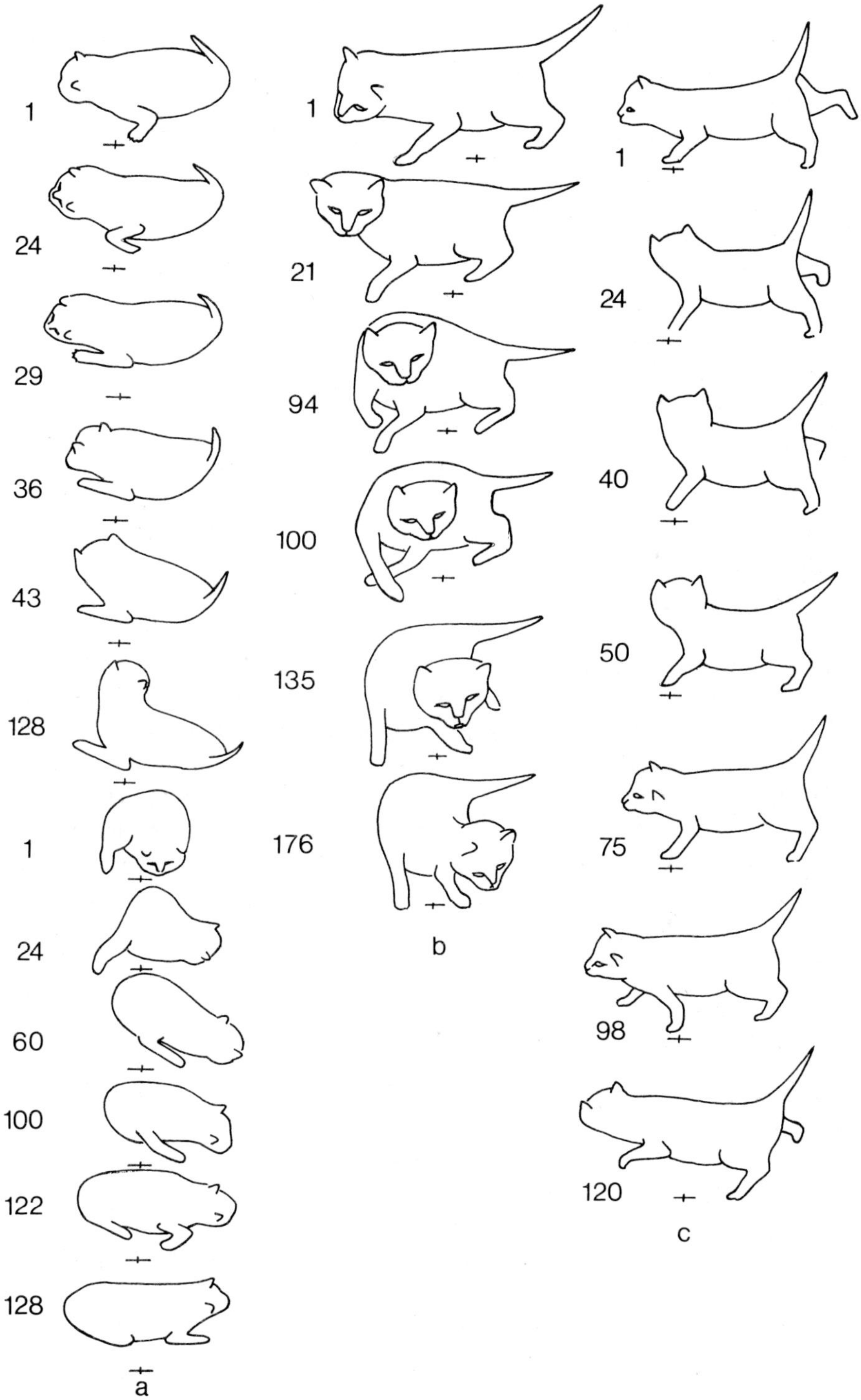
1
24
29
36
43
128
1
24
60
100
122
128
a
1
21
94
100
135
176
b
1
24
40
50
75
98
120
c

muscles show an alternate pattern of stretching and bending in correspondance with the contractions and relaxations of the body musculature. These movements of the extremities as well as that of the body may start without any noticeable influence of the external situation, which argues for an automatic nature of their rhythmicity (Von Holst, 1939). On a smooth surface swaying does not lead to a displacement of the kitten but on a rough surface (e.g. a blanket) a forward locomotion may result from it, especially when the not yet retractible nails from time to time get hooked in the substrate of the fur of the mother. Under such circumstances the legs may assist the kitten in pushing itself up against the belly of the mother (Rosenblatt and Schneirla, 1962).

As long as they are not fed newborn young keep restlessly moving for hours and show an astonishing toughness. In a litter where the mother was not accepting the newborns to suckle they were active for 24 hours before they died. Usually, however, the mother accepts the young and gives them the opportunity to find a nipple. Prechtl and Schleidt (1950, 1951) and Prechtl (1952) have called the swaying movement ''Suchautomatie'' (searching automatism). According to them the frequency of the rhythmic swaying is about 15 swayings per minute, continued for periods of 2-3 minutes, alternated with rest periods of different length. According to Prechtl (1952) the locomotion resulting from swaying would often lead to circular routes ('Kreisen'). We agree that the routes are not linear and that because of the curvings the nidiculous kitten is not likely to distance itself from the nest area. We have occasionally seen this circling in young kittens, and when in contact with a littermate mutual circling may result, but – in our opinion – naming this locomotion 'circling' ('Kreisen') wrongly suggests a considerable amount of regularity.

The licking by the mother affects swaying in two ways. When she presses the tongue against the body and brushes it over a distance of 2 to 3 inches on one side of the spine, the young reacts by contracting its muscles on that side (an effect which can also be brought about experimentally by touching the kitten with the finger). As a result the kitten exposes the contralateral side which is then licked and consequently contracted. In this way the automatic swaying movement is externally induced and moreover a directive effect is added because the mother usually begins her brushing near the head, the kitten moving forward towards the mother. If a more powerful stimulus than licking is applied to the head region the kitten does not move towards it but away, thus avoiding the stimulus source (Schneirla, 1959).

As soon as the kitten has reached the body of the mother, it shoves its chin on her flank. Meanwhile the head continues sideways movements, though with smaller amplitude, while the mouth is opened. The chin pushes a slight depression in the abdomen of the mother and consequently when during the jerky movement it meets a teat the latter is bent into the direction of the mouth and easily slips into it. Thereupon the mouth is closed and the nipple firmly held. Already when the area around the teat is touched a droplet of milk often appears.

The kitten always keeps the chin in contact with the abdomen of the mother, not the upper parts of its head or nose. At this age it never pushes its head under the mothers body, as nidifugous mammals do (Prechtl and Schleidt, 1951). The same behaviour was seen when the mother was replaced by a dummy, such as a hotwater bottle wrapped in cloth. Young left alone by the mother make swaying searching movements and mew when they are awake; they only come to rest when they have managed to put their chin on the body of a littermate. Prechtl (1952) states that at the end of a period of swaying the snout would be "awkwardly bored" ("ein fahriges Boren") into the substrate. Prechtl and Schleidt (1951) report that the kitten directs this movement towards horizontal slits, but not to vertical ones presented in a vertical barrier on a horizontal substrate. However, they neither describe the experimental situation nor the movements precisely enough for us to decide whether or not some disagreement exists between our observations and theirs.

Kittens which have drunk well so that their bellies are swollen, are less able to move around by swaying than newborns with an empty stomach, because their legs are extended out sideways, hardly touching the ground (Pl. III[a]). After 5 days however, the shoulders, pelvic girdles, and the legs have developed so far that the legs are kept under – and no longer besides – the body of the young animal. Also at this age the legs are not yet capable of giving support, but the new position of the legs gives the kitten more grip on the substrate. The body is still only kept horizontal, resting on the belly, while each body movement is still initiated by movement of the head. After a week the weight of the kitten is about doubled.

At the age of 12 days our kittens could sit with the longitudinal axis of the body in a diagonal position, actively supported by their legs. In this position the elbow and the wrist of the forelegs were free from the substrate, on which the toes stood with retracted claws. The feet of the hindlegs were in touch with the substrate as far as the heels. Rolling over was no longer exclusively executed by movements of the spine, the kittens could now use a hind leg as a lever. The crawling movements made at this age are pictured in Fig. 1[b].

At 17 days the kittens can stand on all fours. Walking, however, is awkward because the kitten does not yet manage to adjust the shifts of balance occurring when locomotion in initiated by a movement of the head following by the body and by an appropriate lifting and displacement of the legs. However, this rapidly improves during the next days.

At 21 days a sitting kitten can lift one foreleg and move it freely in the air. It can also scratch its head with a hindleg, but slight disturbances still make the kittens roll over. Fig. 1[c] gives film frames of walking at this age.

At 23 days a kitten, sitting on its hind legs can move both forelegs alternately in the air. It can push itself up on its hindlegs and stand on the toes, but it still easily falls back on the full foot and – as judged from the occurrence of trembling – it is readily fatigued. Depending on height and material of the nestbox, it can climb out of it and stumble around in a wider area. It can drink from a trough and lick on solid food. It can squat down for urinating and defecating and it can bury urine and feces by directed scratching with one frontleg. Most of the time, however, the mother still licks up the urine and although this behaviour becomes gradually less common when the young grow older, it can still be observed in the second month. The need to urinate is signalled to the mother by loud mewing; moreover the kitten evidently searches for a suitable place. When the mother approaches, the kitten often presents itself by rolling on its back.

In order to put its chin on the belly of the mother the newborn must contract part of its dorsal musculature. Action of the dorsoventral musculature is also necessary for raising the body in sitting and, in a more complicated sequence, in rising on the hind legs and pulling itself up by the forelegs to climb out of the nestbox. Dorsoventral curving of the spine becomes increasingly frequent as a part of various behaviour patterns after 4 weeks. For instance, when a 4-week old kitten stepped from a dry surface on to wet grass and irritated its ventral side, it rose on its hind legs, curving its spine and proceeded with undulating jumping movements. The dorso-ventral musculature is also involved at an older stage in galloping, when the hind feet are pulled up until they are below the shoulders, whereupon the forelegs are thrown forward. This pattern is seen when the cat is over two months old (Fig. 2).

When awaking an almost 6-week old kitten, after yawning (with the point of the tongue curled upwards) and while laying on its ventral side, may stretch the forelegs and progressively curve its spine upward to finally stretching its hindlegs and tail in a diagonally upward and hindward position. Subsequently the kitten moves its body forward, circling around the stretched hindlegs, and by stepping forward with the hindlegs it reaches a normal position.

This description of the gradual development of the locomotory patterns thus shows a succession of three phases. The first phase is marked by the exclusive domination of the body musculature producing the swaying pattern which is lead by the head. This pattern occurs autonomously as well as reflexively, whereas the legs either passively follow the body movements or stretch or bend as a result of external stimulation.

In the second phase the body musculature still dominates, but the legs are used to support the body in particular postures, such as crouching, sitting, standing or walking. In these cases the curving body brings the legs into a suitable position in which, when receiving the necessary stimulation, they can react appropriately for supporting the body and maintaining equilibrium. The number of movements the legs can make has increased, but their acitivity is secondary to that of the body.

In the third phase the legs have become more emancipated. To a certain extent they can move independently of the body, sometimes oriented with respect to a stimulus source such as in scratching, or sometimes without an apparent external orienting cue.

During the first two weeks we only saw our kittens *approach* a stimulus source (i.e. the mother). Thereafter, when the mother was more often away from the nestbox, we could evoke *retreat* by moving a novel object rapidly towards the young. During approaching head and body are moved forward, the

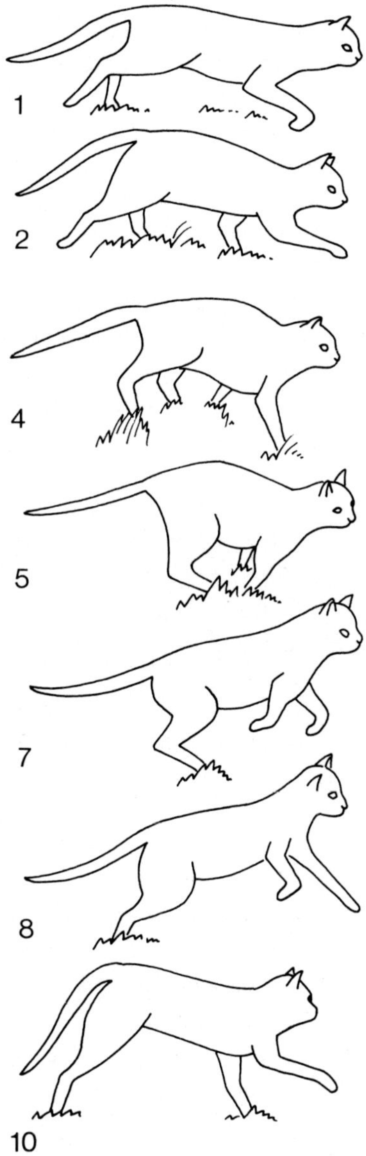

Fig. 2. Five month old. Running. The swaying of the head followed by the sigmoid movement of the body is noticeable but now all movements proceed much more fluently. The alternate movement of the forelegs is more distinct than that of the hindlegs. The spine is regularly curved dorsoventrally (frames 4 and 5) and stretched when the hindlegs push off (frames 7 and 8).

auricles are erected and maximally directed towards the object approached, the eyes are wide open and the whiskers form a frontally directed concave screen.

The withdrawal of the head and body from a stimulus first becomes possible in the second phase (3rd week), when the legs have acquired some independence. The retreating movement becomes more elaborate with the growing capacities of the eyes. In retreat the auricles are directed sideaways and eventually folded and laid down, the eyes are closed to slits and the screen of whiskers is folded in two bundles which are moved backwards and laid along the cheek. From its earliest appearance retreat can be combined with hissing and spitting. These latter activities however, could already be distinguished on the 2nd day in the nest when the mother was away and a relatively strong tactile or olfactory stimulus was given to the kittens. Hissing and spitting seem part of the same pattern, spitting usually following some successive hisses. The mouth is open, the upper lips are curled up and the tongue is kept flat and clearly visible (Pl. II[f]).

Retreat begins with a backward and downward movement of the head, accompanied with curving of the spine and an upward movement of the hind parts. Initially the legs remain in place, the forelegs bending at the elbow take up an oblique position. The pattern can be performed while sitting and while standing; in the latter case the hindlegs are kept vertical and stretched, the spine is curved, and a first indication of the arch is present. In the sitting position the kitten may crawl backward by alternating movements of the legs. When continuing its retreat from a standing position a kitten usually first moves sidewards and then turns to run away (Pl. V[d]).

From the beginning sucking kittens often show a distinct milk-treading. The forelegs laying laterally to the head on the belly of the mother alternately bend and stretch from the elbow to the claws. After the 2nd week when a leg is stretched the phalangi and claws are fully extended, then touch the mothers skin, and while the claws are gradually retracted and the fingers and wrist bent the paw brushes through the fur over a maximal distance equal to that of the length of the leg below the elbow. When the foreleg is maximally bent it is lifted and, when free from the belly, brought back to the original stretched position on the skin. One stroke takes circa 2 seconds; the bouts of strokes may last several minutes. During milk-treading the kittens always make a loud purring sound. The treading and scraping movements of the mother cat mentioned above (p. 11) resemble milk-treading with respect to the pattern of muscle contractions, relative amplitude, and frequency.

The so called "claw sharpening" (Leyhausen, 1956[b]) seems to consist of the same leg coordination as is used in milk-treading. The cat places its forelegs on a rough substrate, like the bark of a tree (which may make all possible angles with the ground) and performs the first phase (leaning backward) of the stretching movement (p. 18) with the treading pattern superimposed. The amplitude of the alternating leg movements tends to be smaller in this case, probably because of the resistance met by the claws which scratch through the substrate.

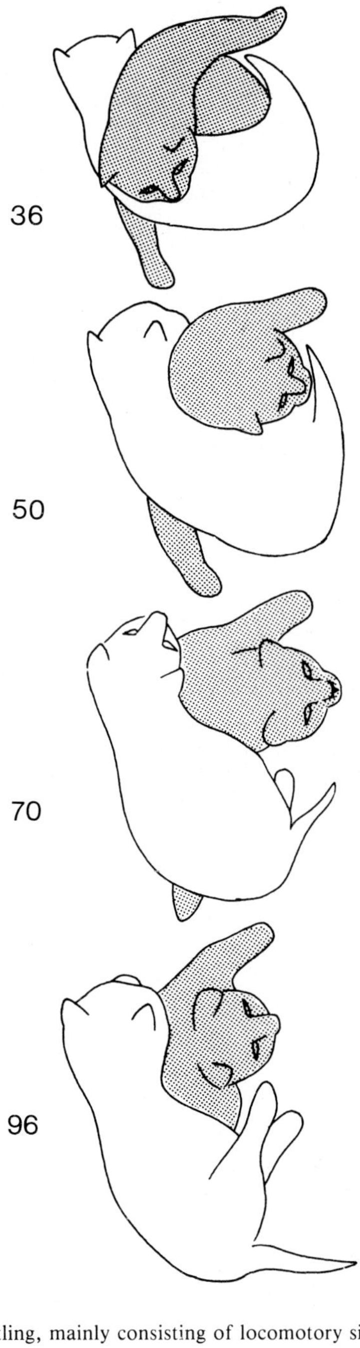

Fig. 3. Two weeks old. Wrestling, mainly consisting of locomotory sigmoid mutual pushing with the bodies.

It has been stated above that kittens left alone by the mother come to rest when they manage to put their chins on the body of a littermate. When this is done mutually by the kittens the entire litter comes to rest in a cluster. However, as a consequence of this behaviour a kitten easily gets into an uncomfortable position and this happens more often when the kittens grow older. At about 20 days kittens which were under a littermate were seen first to retreat by crawling backward and, when free, to approach again pushing up the head and trying to shove the chin on the littermate. The legs follow the body movement and because they are sometimes lifted we must assume that at least part of the leg movement does not just occur passively. Kittens a few days older were seen to work themselves on top of a littermate by pushing themselves on the hindlegs with clearly directed alternated slamming movements of the free

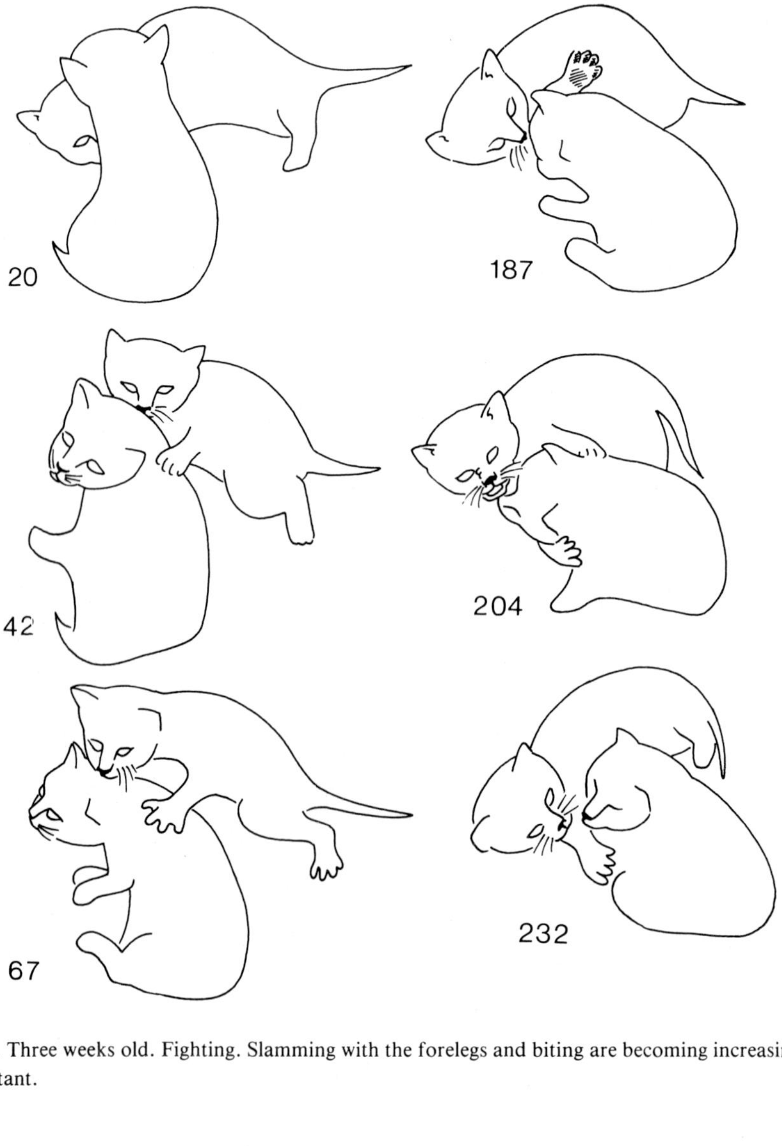

Fig. 4. Three weeks old. Fighting. Slamming with the forelegs and biting are becoming increasingly important.

forelegs, phalangi and nails extended, and finally come down on the other kitten with mouth open, closing it for a bite. During opening of the mouth the eyes are closed to slits; when the mouth is closed the eyes are opened again (Figs. 3 and 4).

The resemblance of this obvious attack with the just described position correction behaviour suggests that both patterns might be ontogenetical stages of wrestling behaviour.

The attack movements become increasingly frequent during the following days and – just as in locomotion – the role of the legs becomes more and more prominent. At 30 days an upward jump can be made by suddenly stretching the hindlegs. Then a kitten may attack a littermate by jumping at it from some distance. When during its attack the hindlegs of a kitten loose touch with the substrate (e.g. because the attacked littermate rolls over) a characteristic alternating kicking movement with the hindlegs, directed at the body of the opponent is often performed (Figs. 5 and 6).

For defense the attacked kitten has essentially the same movements at its disposal. The main difference between attack and defense is in its orientation. In the extreme form of defense the kitten rolls on its back and, depending on the position of the attacker, bites, slams with the forelegs, and/or kicks with the hindlegs (Pl. IV[e]). Particularly the latter movement is effective in defense because the defending kitten, while pushing itself off with the back against the floor, can move the hindlegs freely – without competition with the locomotory function – to kick away the attacker (Fig. 6).

Despite the occurrence of agonistic behaviour the tendency to stay quietly in body contact with littermates, which is so marked during the first three weeks, remains thereafter; the kittens prefer to sleep in clusters. When awakening they often lick each other. As soon as the kittens begin to walk about a meeting ceremony develops (Pl. V[b]). Encountering each other frontally (an aggressor usually attacks from behind) two kittens lightly touch noses, and either one of them or both advance, touching each others sides while passing. The tail is raised and bent over the other kitten's body. At high intensity the cheek is rubbed along the cheek and the flank of the other and the head may repeatedly be rubbed at other places on the body of the partner. Eventually they mutually nose at the anal areas.

Rolling on the back, as mentioned in the removal of urine by the mother and in defense against attack from a littermate, becomes gradually used as a pattern to invite the approach of another kitten, a potential mate or even its keeper.

Conclusion. During the first four weeks the behaviour patterns distinguished develop progressively, essentially by the addition of new elements. After this period changing continues, but in our opinion these changes have to be considered as adaptive modifications and combinations of the existing patterns; no essentially new patterns appear.

For the understanding of the causation of these changes some knowledge of the sensory capacities is essential. Therefore the development of these sensory

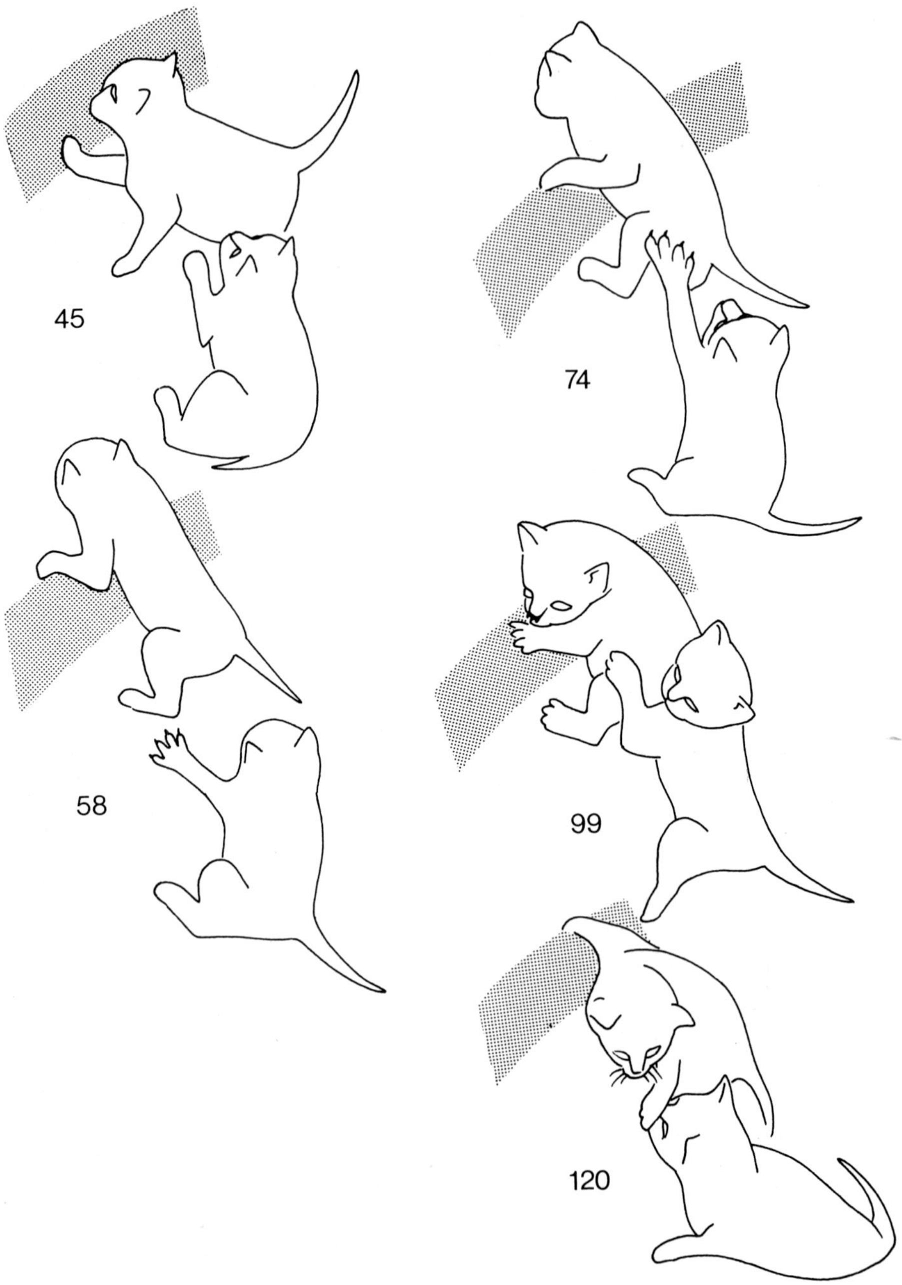

Fig. 5. Five weeks old. Fighting. The lower kitten attacks the upper one while the latter is climbing on the edge of the nest-box. The sequence can be seen in the lower kitten pushing up in the hindlegs and slamming with the left frontleg. In frame 120 the upper kitten retaliates while the lower one begins to assume the defensive position, turning over on its back.

Fig. 6. Five weeks old. Defense. The lower kitten lays on its back, slams at the attacker with a foreleg (frame 10) and pushes it away with the hindlegs (frames 20 and 22).

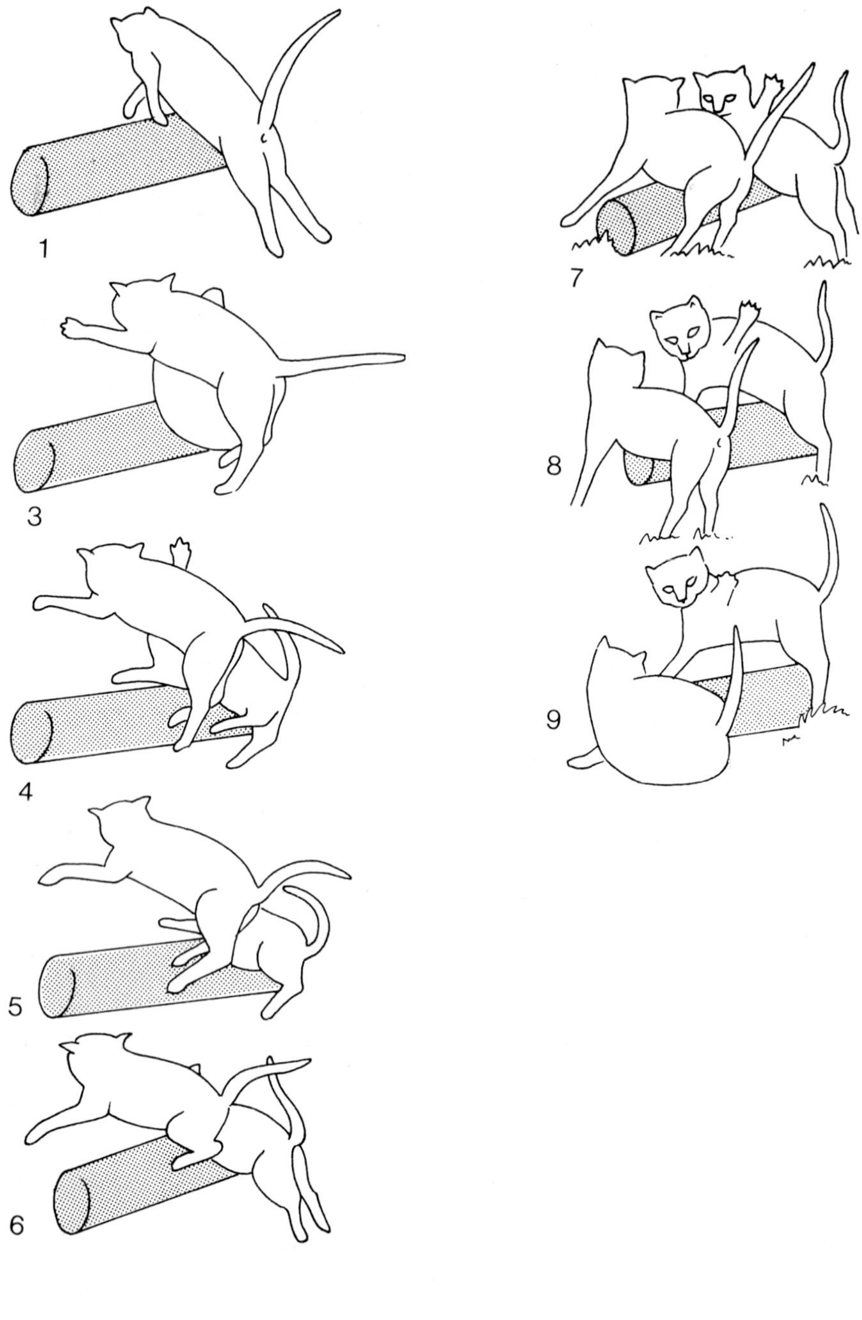

Fig. 7. Seven weeks old. Agonistic encounter with threat. While jumping over a treestem (frame 1) a kitten is suddenly attacked by littermate (frame 3). It turns itself towards the attacker (frames 5 and 6) and both kittens assume the arch posture in lateral orientation (frames 7 and 8). In frame 7, 8 and 9 the attacking kitten slams while the opponent takes up a defensive position.

capacities will be dealt with in the next section, before resuming in section 4 our description of the developing behaviour patterns.

II.3. THE DEVELOPMENT OF THE SENSORY CAPACITIES AND THE ORIENTATION OF THE YOUNG

Tactile, thermal and olfactory senses. In the above we have stated that the newborn kittens, when touched by the mother, react by directed head movements and contractions of the body musculature. Theoretically this touching might involve an adequate stimulus for a tactile, a thermal, or, when the head is touched, also an olfactory sensory organ. The role of temperature was tested by placing a kitten not older than 8 days between two rubber hotwater-bottles wrapped in cloth. One bottle had an outside temperature of 15°C, the other of 25°C. The distance between the bottles matched the width of the kitten, so that while laying still and while swaying it was in contact with both. The behaviour of eight different kittens was watched for 10 minutes each day of the first week. They crawled on both bottles but showed no discrimination between them, neither in first choice nor in time spent on each bottle. This makes it unlikely that the body-temperature of the mother (25°C in the fur at a room temperature of 18°C) has a distinct directive effect on the movement of the newborn or that it could act as a consummatory stimulus to stop searching. For the latter, contact with the nipple is not essential; kittens also come to rest when their chin is on top of some body part of a littermate.

Experiments of Prechtl and Schleidt (1951) gave results which are in agreement with the latter but in contradiction with the former conclusion. These authors report clearly directed movements towards the higher temperature in a similar choice arrangement. However, they give no information on the temperatures actually used; perhaps kittens only respond differentially if the temperature difference exceeds 10°C.

In similar experiments kittens were given a choice between two equally warm hotwater-bottles, one of which was wrapped in a freshly washed cloth and the other in a similar cloth on which the litter and the mother had been laying in the nestbox for 24 hours. Again no discrimination was found; this result is in agreement with Prechtl and Schleidt (1951). The same was true when a bottle wrapped in fur was given against one wrapped in cloth. In contrast, when adult cats were given a choice between these two materials they preferred to lay on the fur. These results argue against a role of chemical and specific tactile stimuli in orienting the approach towards the nipple.

The negative result does not exclude the role of olfactory stimuli on orientation as a whole at this age. On the contrary, Rosenblatt, Turkewitz and Schneirla (1969) have demonstrated the influence of olfaction in young kittens displaced to different areas of the cage to find their way back to the home (= nest) corner.

The behaviour by which the nipple is found seems feasible with tactile orientation only; however our observations and experiments contain no arguments to exclude the role of chemical stimuli. We have clear but occasional evidence of the influence of olfaction on the localisation of food when, after a piece of fish had been put outside the nestbox, 25-day old kittens (which had no experience with fish) climbed out and began to search in circles (see p. 70). Having found the food, they sniffed at it without eating (these kittens did not eat solid food before they were 28 days old).

Eyes. At birth the eyes are closed. They open between the 7th and the 10th day. In an attempt to investigate at which age the young would be able to see we have, from the day on which the eyes opened, tried daily in a great number of kittens to elicit reactions either from the eyeballs or from the head to objects moved in front of the head. Clear reactions were not obtained before the age of 15 days. Thereafter, a kitten laying on its back while the mother was licking it, might move its frontlegs accurately directed towards her nose, at a distance of 5 cm from the kittens eyes. One has to realise, however, that the failure to obtain earlier behavioural reactions to visual stimuli may not be due to an insufficient development of vision but of the motor apparatus involved.

At the age of 25 days the kittens were able to approach and to follow with movements of the head objects moving slowly (e.g. a walking mouse) in front of them at distances within 30 cm. They could not yet keep track of a rapidly moving object (e.g. a running mouse).

Our impression is that from this age on the motor and not the visual capacities are limiting in determining the rate of further development.

Ears. At birth the external auditory canal appears closed, the bases of the pinna being pressed against the proximal wall of the canal. The plane of the pinna makes an almost vertical angle with the surface of the skull Pl. II[a]).

After one week the ear split opens gradually (Pl. II[b]). After the 6th week the auricles are full-grown and, as the skull is smaller then the skull of the adult, the ears of a kitten are impressively large at that age. (Pl. III[e]).

From the first day on we have looked for reactions to noises in all kittens. The earliest reaction we noticed was a twitching of the pinna and this we did not observe before the kittens were 15 days old. They reacted with "mewing" to the "mrauw" calls (p. 36) of the mother and to various noises we made.

At 25 days the ears could quickly turn to the source of a noise. Noises alarmed kittens in rest and could also distract them from visually following an object.

Conclusion. When the kittens are 28 days old all sensory organs are fully functioning and are used in coordination with each other. When a noise is heard, the source is sought first with the eyes through turning of the head, then with the nose by approaching with the head and when reached, touched with a frontpaw and possibly taken up in the mouth. However, at this age attention to a definite source of stimulation is usually lost within about 2 minutes.

II.4. THE FURTHER DEVELOPMENT OF MOTOR BEHAVIOUR AFTER LEAVING THE NEST

Comparison of the behaviour of more than a hundred kittens from circa 30 litters showed that at the end of the first month very little difference existed between individuals in the number of different behaviour patterns available to them and in the stages of development these patterns had reached. In the course of the next month however, individual variations became distinctly apparent. Since this variability developed during the period in which the kittens were roaming around in the outdoor pens or free in the garden, the question rose whether this individual variability was caused by different experience. In this period a behaviour which is commonly called "play" occurs very frequently. This leads to the question which role this "play" might have in obtaining such experience. Finally, it should also be asked which causal mechanisms are underlying the phenomenon of "play".

We shall first aproach the answer of these questions by describing carefully, based on our film material, the kind of changes occurring in the various behaviour patterns. In Chapter III we shall give for one functional group of patterns, prey catching, experimental evidence for the role of learning in the causation of individual variability.

Except for resting periods the kittens when outside the nestbox were constantly exploring the environment with their ears and eyes, and by sniffing and manipulating with forepaws and mouth. As, however, the kittens extended this exploratory area only very gradually and as the area around the nestbox was rather monotonous, littermates were during the 5th and 6th week still the most conspicious elements in the environment of a kitten. As long as they were awake the littermates attacked each other mutually with the same behaviour elements as seen in the nestbox, but now the fighting pattern became interspersed and combined with locomotion: walking, running and jumping towards and away from the opponent (Figs. 4 and 5). The moving littermate appeared a stronger stimulus for attack than the rather passive one in the nestbox; the free running kittens often tried to catch one another, frequently switching the role of attacker and attacked (Fig. 7). During this activity they were increasingly exploring features of the environment and performing activities like approaching under cover, laying in ambush, avoiding, escaping, and looking for places to rest. The kittens began to show individual preferences with regard to types of activity and particular routes or spots in the area.

When the kittens are 4 weeks old their bites do not yet hurt, but in the following weeks bites become stronger and really painful; a kitten bitten at that time by a littermate may scream in pain. In the 5th week also the nails begin to scratch more effectively. This development of the weapons is very soon followed by the first signs of respect for each others attack. The attacker no longer completes its jumps; it inhibits its movement in the very last phase. At about 33 days this invariably leads to a loss of equilibrium, the attacking kitten awkwardly falling back. A few days later, however, the kitten merely makes the intention movement of jumping, without losing its equilibrium. It is now able to inhibit a galloping jump shortly before the opponent is reached, at that moment the hind feet are placed forward, by straightening the frontlegs and keeping them on the ground instead of throwing them up again. Consistent with the short distance between the front and hind feet the spine is strongly curved upward. After stopping the tail continues the forward movement and is consequently pointing upward and bent over the hump. Subsequently, keeping its hindlegs on the ground, the kitten lifts the frontpart of its body and alternately sways its frontlegs through the air, the nails extended and with clappering teeth (*vertical stance*, West, 1974; *rear*, Barrett and Bateson, 1978). After this display in front of the opponent the forelegs fall back again, usually only for a short period as the kittens tends to repeat the display several times. During this phase the pinnae, which during the approach were directed forwards, are turned sidewards.

One may say that when the kitten performs an arched display the body part which is nearest to its opponent begins to withdraw (as can be deduced from the position of the pinnae, whiskers en eyelids) while the hind part is still approaching. The opposite tendencies often lead to a sideways orientation of the display, a compromise between the orientation of withdrawal and approach. Kruijt (1964) has earlier given a similar explanation for the side display and waltzing in the red jungle fowl.

In the 6th week this display becomes reduced to a series of successive up and down movements of the frontal part of the body following a sudden stop after a rushing approach. The forelegs move mainly up and down with the body; the sideways swaying movements have become reduced (Fig. 8).

In the 8th week these sham-attacks become further reduced. In the same situation the body is only slightly lifted. The forefeet are now kept on the ground and while the forelegs remain fully stretched and "frozen", the hind legs make one more step forward. The spine assumes a curve, starting in the shoulders and flowing backwards towards the loins. The fur is raised and the thus thickened tail is erected. The pinnae tend to be turned backwards. If the opponent dodges sideways the initiator stifly turns its head (*neckflex*), following the opponent with the eyes (Pl. IV[a,b]). When the situation relaxes the tense posture is gradually "unfolded" and the kittens stifly distanciate themselves from each other. The same posture may be assumed by a kitten slowly retreating from an attack. This time the curvature results from the forelegs retreating while the hindlegs are staying put.

Whereas during the period of development the posture seems to be shaped by the situation in the way described, at a later stage it can be produced very quickly and at sudden and unexpected encounters. The complete performance has been named *"Katzenbuckel"* by Leyhausen (1956[a]) and *arch*, combined or not with *neckflex* by Barrett and Bateson (1978).

When during this period when a kitten strikes at its opponent it also first extends fingers and nails, but shortly before it hits, the claws are retracted and the fingers bent, through which the effect of the paw is softened. Still later (between week 12 and 16), striking becomes reduced to sweeping a paw closely in front of the nose of the adversery. The latter reacts by closing the eyes and moving the pinnae sidewards, but only for a short moment. During this very inhibited threat both kittens may slowly wave their tails.

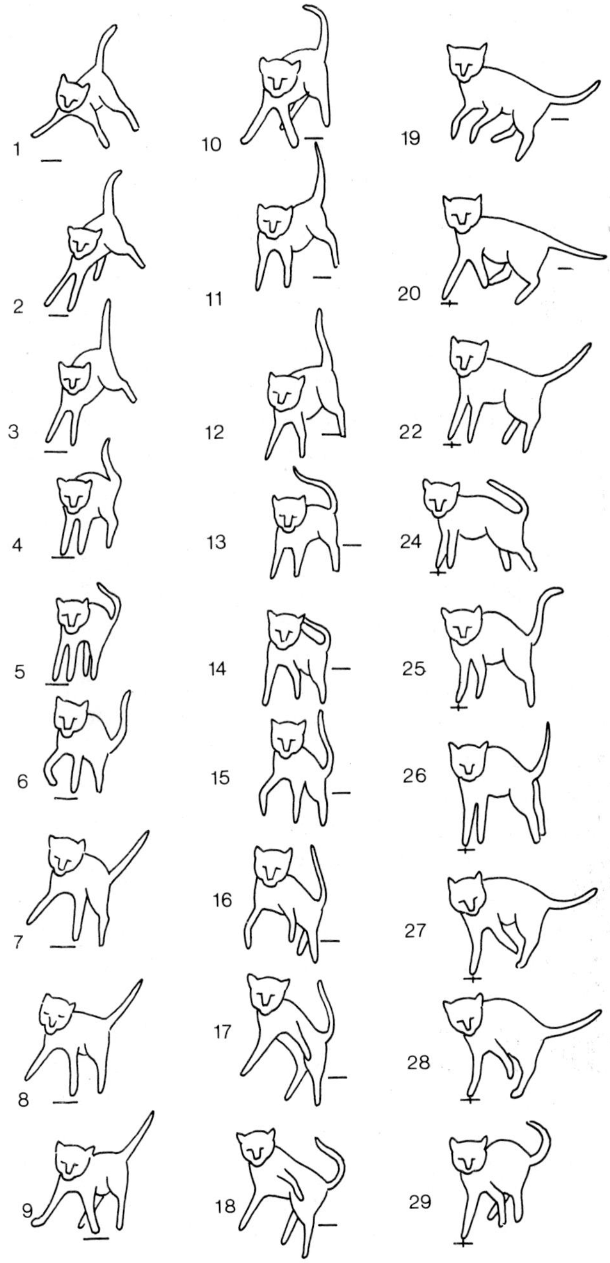

Fig. 8. Seven weeks old. Agonistic behaviour of a kitten, while remaining almost on one spot after approaching an opponent (not visible). Frames 6 to 10 and 15-19 show the alternate slamming with the frontpaws, during which in frame 6, 17 and 18 the body is raised on the hindlegs (rear, vertical stance). In the frames 5, 13 and 24-26 can be seen how the forward body movement is checked by the positioned frontlegs while the arch is being formed.

Between the 4th and 6th week a kitten is usually attacked with a rushing approach as described above. At the end of this period, however, running was more and more replaced by stalking. In stalking the kitten makes use of available cover between itself and the kitten to be attacked. Apart from the low speed, the movement is characterised by the body posture. During the movement all legs, while slowly making the normal locomotion, are strongly bent at the elbow and wrist and at the knee and the heel joints, so that the ventral side nearly touches the ground. The pinnae are erected and the eyes wide open (Pl. V[e]).

This proceeding can be interrupted by the so called "Wackeln" (Leyhausen, 1956[a], vacillating) during which the kitten stops with hind legs relatively far forward and the hind part of the back showing a slight hump. In this position, beginning at the shoulders, the kitten makes with its entire body a lateral quivering movement. Only the head is kept still, the tail making a slow waving movement with greater amplitude than in the normal walking situation.'

After this vacillating the animal either jumps at its target or resumes stalking. The waving of the tail (here as well as in the cases just mentioned above) might be considered as an inhibited forward locomotion. The lateral quivering movements which have a much smaller amplitude might similarly result from ambivalence. The depressed body posture might be a third and the usually slow pace a fourth indication suggesting that stalking is brought about by inhibition of locomotion.

When a kitten gets hold of a littermate after a rush, or after a jump preceded by stalking, both may roll over gripping one another with the frontlegs and pushing with the hindlegs, while biting may take place.

As long as the kittens are orientated belly to belly, alternately changing roles as attacker and attacked, this behaviour can appropriately be called wrestling (Pl. IV[e]). However, in the course of this activity one kitten may grasp the opponent by the nape; then the behaviour which follows very much resembles a predatorial attack, but also a copulation attempt. Resemblance with copulation is even more striking when the acting kitten tries to place a hindleg on each side of the back of the other one. Such mounting attempts were more often, but not exclusively carried out by tomcats, but before the 6th month penis erection in the male or sideways bending of the tail in the female were never observed. The dorsally attacked kitten usually stays extremely quiet, relaxing its musculature. As a rule it is soon released, after which the kittens start licking each other.

The free-roaming kittens may approach the mother with all forms of agonistic behaviour described above for attack on littermates. The mother may counter-attack just like a littermate, but more often, while slowly waving her tail, she places a leg over the kitten and starts licking it, whereupon the kitten starts suckling.

The littermates and the mother were the most conspicuous objects moving in the environment of the kitten in close proximity of the nest box and more or less independently of its own activity. When the area of exploration was gradually enlarged from the 6th week on small animals, in particular insects, were often encountered. The activities released by such objects are essentially the same as shown towards other kittens, but adjusted to their relatively small sizes. Sometimes objects, which could only move passively, may attract attention and can subsequently be kicked and thrown with the paws and the mouth. This behaviour was studied (with the help of cinéfilm) by presenting to the kittens a rubber ball (diam. 3 cm) swinging on a string 10 cm above them.

Novel immovable objects also drew the attention of the exploring kittens. Six-week old kittens, for instance, ran towards the stems of big trees and climbed upward with walking movements of the hind legs and alternate movements of the sideways spread forelegs. They showed no difficulty in climbing upwards but hesitated in coming down. At their first attempt they now and then interrupted climbing, looked downward and showed intentions to descend, but often after some indications of being frightened, climbed upwards again. When they finally decided to come down, they tried it head downwards and invariably fell of the tree. After a few trials, however, they only used this way of descending when near the ground, otherwise they came down head upwards, using the behaviour pattern of ascending reversely.

When descending in a tree with low hanging, easily waving branches, the kittens first jumped downwards from branch to branch, randomly, rapidly and clumsily, but already after a second attempt they managed to balance quietly and in harmony with the waving cadence of the branches, waiting for the proper time to jump.

The rapidity with which locomotory and manipulatory activities were adjusted to the objects involved was striking. Ofting an improvement was visible after only one trial.

At the age of 12 weeks marked individual differences could be observed between kittens of different litters and also between individual kittens, with regard to their behavioural skills and their possibilities to cope with different situations. Many of the differences could be correlated with differences in environment the kittens had experienced. For instance kittens of a litter kept indoors could walk on a slippery floor, and kittens kept outdoors could climb and descend trees whereas each behaved awkwardly in the unfamiliar situation. Kittens of the same litter tended to assume individual ways (tricks) to leave or enter the house, beg for food etc. All these observations are casual evidence that during the 4th-12th weeks as a result of the behaviour of the kitten a great number of important learning processes take place which adapt it to its species-specific "Umwelt" (Von Uexküll, 1909) as well as to its individual-specific environment. Systematic observations and experiments on how this adjustment takes place have been concentrated on one aspect: prey-catching behaviour, and will be dealt with in Chapter III.

II.5. CONCLUSIONS

During the first four weeks of life the motor apparatus, the coordinations by which this apparatus is operated in several functional contexts, and the sensory organs develop up to their maximal capacities. In the next four weeks the activities run gradually more smoothly, they become better oriented and the rate at which the behaviour in general proceeds tends to quicken. The changes, modifications and novelties in the form of activities can be understood as the consequence of counteracting inhibiting tendencies. In the agonistic encounters of kittens this leads to threat behaviour, in locomotion to stalking, and in the manipulations with objects from the environment to adequate timing of the activities and an appropriate dosage of their force. The inhibition seems to act on the various aspects of a pattern in a differentiated way. For instance, when during striking with the forepaw the nails are already fully retracted and the fingers bent, the paw as a whole may still be forcefully brought forward; also the stopping of a run becomes evident earlier in the front than in the hindlegs. The kind and degree of inhibition was also found to be differentiated with regard to the object concerned. Attack behaviour, for instance, became inhibited towards littermates or other familiar cats, however without the cats losing their attacking potentialities in other occasions. It is understandable that the differentiated inhibition of attack only becomes possible after sufficient development of the sensory organs.

The interaction between the attack behaviour and the inhibiting influences does not lead to a number of discrete threat activities, each corresponding with a certain balance between the tendency to attack and to escape, as for instance Moynihan (1955) distinguished in the black-headed gull and between which transitory forms of threat behaviour are reported to be rare (Tinbergen, 1959). On the contrary a grading scale between the most extreme form of attack and of retreat can be seen, as has already been pointed out by Leyhausen (1956[a]) for the body postures and facial expressions during agonistic behaviour.

The influence of the inhibitions on the original behaviour is highly important for the social behaviour of the cats. It is important in various aspects of the phenomenon called "socialization". Moreover, inhibitions provide a safequard for the animal in all kinds of instrumental behaviour, e.g. in prey-catching, manipulatory exploration, etc. Young kittens in general behave less carefully than older ones.

The facts (1) that the influence of inhibition starts after a considerable amount of perception had become possible, (2) that differentiation occurs with regard to the object concerned and (3) that individual differences in behaviour become gradually more prominent after the 4th week, suggest that learning may have an important influence in the developmental process. The occurrence of learning before the 4th week has been demonstrated by Ewer (1960) for nipple orientation and nipple preference of kittens as early as in the first week. Rosenblatt, Turkewitz and Schneirla (1959), 1969) have shown early learning to be involved in the direction of movements towards the mother and in spatial orientation inside the cage, respectively.

The following study of the development of prey-catching will give some opportunities to test the role of learning in the ontogeny of species-specific behaviour.

III. Factors involved in the development of prey-catching

III.1. THE ADULT CATCHING PATTERNS

As detailed descriptions of the prey-catching pattern and its variation are given by Leyhausen (1952, 1956[a], 1965) we can suffice here with a brief description relevant for the present study.

Sound and vision are the most important sensory modalities by which a cat detects a prey from a distance. A cat searching for prey walks short distances, interrupted by shorter or longer periods of watching and listening, often while sitting on an elevated observation post. Movements and rustling or high-pitched sounds in particular evoke extreme altertness in the cat. When the stimulus is received again but the source is still at some distance, the cat carefully stalks towards it, using cover as much as possible. When the vegetation is not high, the cat, while moving, keeps its body close to the bottom, now and then waving its tail, particularly when slowing down. When near the potential prey – the absolute distance depending on topographical circumstances – the cat either rushes towards it or jumps and pounces upon it.

The tools used to catch the prey are the mouth and the claws of both frontlegs (see Eisenberg and Leyhausen, 1972). In an easy situation only the mouth may be used; the prey is bitten behind the ears and killed by one or more bites. If a prey is more difficult to handle because it is relatively big, if it is defending itself, if it tries to run away, or if it is partly hidden, the cat uses its paw to hold it. When the paws almost simultaneously reach towards the prey, the fingers are stretched and the claws extended; after the latter have hooked into the prey, the fingers move back, keeping the prey under control until the mouth has got hold of it. In case the cat does not at once succeed in grasping and holding the prey, it will use mutually co-ordinated movements with both frontlegs to bring it under control. Such movements often occur when a bird is caught just above the ground. In our experience the prey is always killed with a bite, not with the manipulations of the forepaws. Just before the cat bites its prey to death, its behaviour becomes more determined. The jaws which so far only held the prey begin to gnaw it. After the prey has been torn, pieces are severed with snipping movements of the jaws, alternately with the left and the right side of the mouth (Pl. VI[f]). When the size of the prey is relatively big in comparison with that of the cat the frontpaws may be used to hold it in place on the ground. Relatively small prey can be cut to pieces while dangling free in the mouth of the cat. Small rodents are as a rule entirely devoured by adult cats (i.e. with hair and bones, leaving no remains).

When attempting to retrieve a prey from a crevice or a narrow corner the cat uses one frontleg with stretched fingers and extended nails to pull it out. In particular the dew-claw plays a role in getting a grip on the prey during such manipulations. This angling movement is also used by cats retrieving fish from shallow water. It was a typical element in the behaviour of the cats in the fishcatching experiments described in III. 2.2.1.2. In this way they avoid getting the nose wet, which obviously strongly disturbs them because it evokes sneezing and thorough wiping with the frontleg. This "angling" is especially efficient when the fish can be cornered against a wall but may even be effective in open shallow water. Only one of our cats used both frontlegs in catching and holding a fish in the water; it then pulled the fish up by raising the frontal part of the body on the hind legs, bringing it to the mouth (see III.2.2.2).

III.2. THE DEVELOPMENT OF THE CATCHING PATTERNS

III.2.1. *Description of the course of development of predatorial behaviour*

When a litter is about four weeks old, the mother starts to carry prey to the kittens. When approaching the nest with a prey, she utters a typical call characterized by its relative loudness, prolonged "au", and the absence of a rolling "r", instead of which a "l" or "w" can be heard (lauw or wauw). This is probably the same sound as the "mrauw", by which in a later stage the mother calls the young when the family is roaming about in the home range, changed through the acoustical effects of carrying a prey in the open mouth.

At this sign the kittens go to meet the mother at a shorter (some decimeters) or longer (several meters) distance from the nest depending on their age. The mother drops the prey (in our observations usually mice) in front of the young. One or more of the young may show interest; they may sniff, lick and bite at the prey and, when the latter is still alive and moving, pursue it. During this pursuit the prey may attack to defend itself, which usually makes the kitten retreat. If none of the kittens kill the prey within a time span varying from some minutes to half an hour the mother catches the prey, kills it and again drops it in front of the kittens. One of them may tear the prey apart or otherwise the mother will begin to eat and allows the kittens to join and dislodge parts of the prey (Pl. VIII). As soon as a kitten has captured a prey or a part of it, littermates and even the mother are no longer tolerated at short distances; at their approach the kitten growls and slams at them with its frontlegs and extended nails. But if the kitten is small it may be satiated before the prey is finished; it then leaves the rests to its littermates.

The earliest age at which killing was observed was 27 days; the prey was an extremely small mouse (12 g). On the average in the laboratory tests kittens killed their first mouse (20 – 25 g) before they were 45 days old.

From the 4th week onwards the kittens increasingly react to noises and moving objects. Insects are followed on the ground and in the air and the kittens try to catch them with the paws when they cannot reach them with their mouth. Particularly butterflies and bumblebees can be zealously hunted, but also leaves and frogs are taken as targets.

At first the time the kittens concentrate on one object is short, but during the 5th and 6th week this time increases. Thereafter the kitten usually holds on to a potential prey until it has either caught the prey or until the latter has managed to escape. During this activity the kittens often show interest in what their littermates are doing. Initially this usually leads to an attack upon the "hunting" littermate, but later on the attention of the kitten shifts to the hunted object, even if it is not allowed to approach closely.

The kittens (and at least one adult cat which was originally raised in the laboratory) which we kept at home often brought in shrews from the garden. The shrews were nearly always alive and often kept the litter busy for a while, since they were not readily killed and repeatedly escaped some distance, showing great skill in hiding in narrow spaces. Shrews are said to be distasteful to cats. A number of times we saw a young cat eat a shrew and vomit within a quarter of an hour; we have never seen experienced cats eating shrews.

The majority of our observations on the development of prey-catching concern the capture of mice.

The first few attempts to kill a mouse once caught varies considerably between individual kittens. However, this variability has decreased after four or five mice have been killed; then the pattern has become stereotyped. Initially the kitten may hold the mouse in every possible direction and bite it repeatedly and in any part of the body (legs, tail, back, head, snout) (Pl. VII^{b}). In further attempts the kitten tries more and more to orient the mouse head forward between its paws and increasingly directs the bites at the area behind the ears. At the age of eight weeks all experienced kittens we observed were able to kill a mouse of 20-30 g with one bite in a stereotyped way. The fact that the species-specific catching and killing movements are practiced in their complete form only after the kitten has gone through a period of great activity (of a type corresponding to the notion of "play"), during which ample opportunity to gain experience with various kinds of external stimuli and with the effects of its own movements seems to be present, suggests that in the development of these behaviour patterns learning is involved. Such learning might comprise the acquisition of knowledge of specific characteristics of the prey, the shaping of the motor patterns of catching and killing, their orientation and the co-ordination of their interaction with other motor patterns.

III.2.2. *Experiments on causal factors in the development of prey-catching*

Two series of experiments were undertaken to study under standard conditions the effects of environmental experience on the development of catching, killing and eating a prey. These series are based on the distinction between an "early period" (from the 4th until about the 8th week) and a "later period" (from the 8th until the 40th week). On the basis of our above mentioned observations and of data in the literature it could be assumed that after the early period experienced kittens are able to kill medium sized mice.

Series I deals with the early period. This series aims at a description of the development of prey catching with time under the standard conditions. It provides insight into the age at which catching and killing can be achieved, the possibilities for learning processes to be applied and the influence of the mother on the development of the predatorial behaviour of her young. In Series IA exclusively mice were given as prey, in Series IB only fish. Consequently the possibility existed that in the course of the tests of these series the kittens became conditioned to one particular kind of prey.

Series II deals with the later period. In most experiments of this series the possibilities of the kittens for acquiring experience had been manipulated in the early period. This series wil allow conclusions on the influence of different kinds of experience with two kinds of prey (mice and fish) as well as the influence of littermates during the early period.

III.2.2.1. Series I: The early period

III.2.2.1.1. *Training on mice as prey (Series IA)*

III.2.2.1.1.1. Methods

The kittens used in the experiments of Series I were kept as litters with their mothers in the laboratory cages under the conditions described on p. 8. Almost every day between 8.00 and 14.00 hour, the kittens of each litter were given a training session (which will further be called "test"). Food was given in the home cages 16 hours before and shortly after the tests. On the morning of the test around 8.00, the kittens were given an opportunity to drink milk from a trough. For practical reasons the sequence of tests could not be started with all kittens at the same age and sometimes the daily presentation had to be left out or the sequence of tests had to be shortened because of lack of suitable prey. These irregularities unfortunately interfered with a statistical treatment of the data.

For the tests the kittens were placed in an experimental cage of 1 m per side. In most cases littermates were tested together and in the presence of the mother; however, the litters n_1 and n_2 were tested without the mother, the kittens of litter *z* without the mother and without littermates. When together the kittens were almost immediately at ease, consequently testing could be started without delay.

After the cats had been introduced into the test cages and seemed to be at ease a living mouse was put into the cage and left there for at least ten minutes. When not eaten it was taken out at the end of the period unless a kitten was continuously carrying it or still eating it. Taking a mouse from a kitten while it was being handled was omitted to avoid negative conditioning processes. If during a test none of the kittens caught and killed the mouse the experiment was discontinued until the next day. If the mouse had been killed a second mouse was introduced into the cage. This was done irrespectively of whether the killer was still eating. However, as soon as a mouse had been completely devoured the successful kitten was gently removed from the test cage. The presentation of mice was continued this way until all kittens had had a chance to pursue a mouse for 10 min. The time a kitten concentrated on the mouse was mostly not longer than 5 min. Thereafter it shifted its attention to other things, such as littermates, the cage wall or rubbage on the floor. This means that the later in a sequence of tests on the same day that a kitten has successfully caught its prey, the longer it had been in the presence of a mouse on that day. Therefore, the latency periods, because they are measured from the beginning of each single test, and not from the beginning of a session with the litter, are biassed towards minimum values. The latency values of kittens who caught a mouse in a later test have been influenced (reduced) by the presence of mice offered earlier in the session and caught by littermates. In our opinion the error introduced by this procedure is not large, since usually only one kitten at a time goes after the prey, the others being otherwise engaged. We have preferred the procedure in which the littermates were together in the test cage over the alternative possibility to expose them singly to the test conditions, because in pilot experiments single kittens had shown a strong and continuous tendency to escape, particularly when under 6 weeks of age. Older kittens usually had very short latency times, irrespective of the rank number of the tests in which they got a chance. They were usually keenly watching for the experimenter to introduce the mouse, whereas with younger kittens the mouse could as a rule be placed in the cage unnoticed.

The greatest difficulty in standardizing the experimental conditions was caused by variations in the behaviour of the prey. The mice used in this series weighed from 15-25 g. We tried to use as much as possible mice from non-aggressive strains, but because we depended on a fluctuating supply of surplus animals from breeding experiments for genetical purposes, mice of the same strains were not always sufficiently available. Moreover, even in the same strain the behaviour of the prey was not independent of that of the kittens. If the kitten hesitated this tended to increase the aggressiveness in the mice, causing it to assume the erect posture while vocalizing and biting the kitten in its legs or cheeks.

III.2.2.1.1.2. Results

A quantitative survey of the results of the training experiments of Series IA is given in Table 1. It involves sequences of tests with 40 different kittens. Some litters were always tested in the presence of the mother, others never (column 12).

To facilitate the survey of the data in Table 1, part of it has been regrouped in Tables 2, 3 and 4.

In Table 2 the data of column 4 of Table 1 have been grouped according to different age categories. It shows that most kittens are between 35 and 47 days when killing their first mouse. As seen from column 3 in Table 1 unfortunately not all kittens could at the same age be confronted with mice. In the tests of the "26-34 days" category only 32 kittens were involved and in those of the "35-41 days" category 36; but from the 42nd day on all 40 kittens were tested. Consequently the figures for the first two categories may, in comparison to the others, be somewhat biased to the low side. Three kittens did not kill a mouse during

Table 1. Survey of relevant data and results of Series IA.

Kitten Nr	Litter	Age 1st mouse seen	Age 1st mouse killed	Number of mice seen before	Number of mice caught before	Number of mice eaten before	Number of mice presented before next kill	Number of mice presented from 1st kill on	Number of mice killed under 56 days	Killing ratio under 56 days (%)	Mother present + or absent − (in test)	Number of young present in test	Sex	Weight ratio 6 weeks / 2 weeks
(1)	(2)	(3)	(4)	(5)	(6)	(7)	(8)	(9)	(10)	(11)	(12)	(13)	(14)	(15)
9	n_1	27	33	4	0	2	1	10	9	90	–	4	♂	1.6
10		27	37	6	0	2	1	7	5	71	–	4	♀	2.3
11		27	27	0	0	0	3	13	10	77	–	4	♀	1.9
12		27	37	3	0	0	1	7	6	86	–	4	♀	1.6
17	n_2	36	40	3	0	0	1	11	9	82	–	4	♂	1.9
18		39	45	3	0	0	2	4	3	75	–	4	♀	1.5
19		39	39	0	0	0	1	10	9	90	–	4	♂	1.6
20		36	41	3	0	0	3	8	4	50	–	4	♀	1.7
75	z	29	30	1	1	0	8	12	2	17	–	1	♂	2.0
76		29	44	12	0	0	–	–	–	–	–	1	♀	1.8
77		29	33	4	2	0	1	8	3	38	–	1	♀	1.9
78		29	35	5	2	0	1	7	6	86	–	1	♀	1.8
79		29	30	2	0	1	4	11	4	36	–	1	♂	1.9
1	mr_1	28	35	6	0	5	1	26	23	88	+	4	♂	–
2		28	41	8	0	7	1	25	20	80	+	4	♂	–
3		28	41	6	0	5	1	25	21	84	+	4	♂	–
4		28	35	5	0	4	1	27	16	59	+	4	♀	–
57	cy_1	27	47	23	9	2	4	14	10	71	+	3	♂	2.1
58		27	–	33	18	3	–	–	–	–	+	3	♂	2.0
59		27	50	26	12	4	5	9	2	22	+	3	♀	2.2
68	cy_2	24	–	34	8	0	–	–	–	–	+	2	♀	2.2
69		24	–	34	11	0	–	–	–	–	+	2	♀	2.1
51	mo_1	32	47	17	0	2	2	7	5	71	+	2	♂	1.9
52		32	48	18	0	3	2	6	5	83	+	2	♂	2.1
60	mo_2	42	47	4	1	1	3	9	6	66	+	4	♀	1.7
61		42	45	3	2	1	5	10	5	50	+	4	♀	1.6
62		42	45	3	0	1	1	10	6	60	+	4	♀	2.0
63		42	45	3	3	3	1	11	10	91	+	4	♀	2.1
70	mo_3	32	42	9	2	1	3	15	10	66	+	4	♂	2.1
71		32	42	10	4	4	1	13	5	38	+	4	♂	2.1
72		32	43	10	1	0	1	15	15	100	+	4	♀	1.9
73		32	43	12	11	9	1	13	10	77	+	4	♀	2.1
53	mi_3	26	48	25	1	1	2	7	5	71	+	4	♂	2.3
54		26	42	16	1	1	7	18	10	55	+	4	♂	2.4
55		26	39	13	3	0	1	17	16	94	+	4	♀	2.2
56		26	54	31	10	1	2	3	2	67	+	4	♀	2.0
64	mi_4	28	40	12	2	1	2	18	13	82	+	4	♂	2.3
65		28	51	21	5	0	2	6	4	67	+	4	♀	2.3
66		28	46	17	3	3	1	11	10	91	+	4	♂	2.2
67		28	39	11	1	0	2	17	9	53	+	4	♂	2.2

Table 2. Age at which first mouse was killed with and without mother present during test. (Abstracted from column 4 in table 1).

Age	26-34 days	35-41 days	42-47 days	48-56 days	Over 56 days	Nr of kittens in test
Mother present	0	7	13	4	3	27
Mother absent	5	6	2	0	0	13
Total	5	13	15	4	3	40

the period the training experiments of Series IA lasted and thus before they were 56 days old.

The probability that after a successful kill the next mouse offered is killed has been expressed in different ways. First, column 8 of Table 1 gives the number of mice presented after the first and before the second kill was made; these data are summarized in Table 3. It follows that 50% of the kittens killed the first

Table 3. Number of mice presented before second mouse was killed, with and without mother present during test. (Abstracted from column 8 in table 1).

Number of mice	1	2	3	4	5	6	7	8	Nr of kittens tested
Mother present	11	7	2	1	2		1		24
Mother absent	7	1	2	1				1	12
Total	18	8	4	2	2		1	1	36

mouse presented after the first kill. Another measure of the likelihood that a kitten which has killed a mouse will also kill the following mice offered can be obtained from the cumulative diagrams given in Fig. 9 for the performance of each individual kitten. This figure shows that after the first success failures are relatively rare.

Finally, a grade for the amount of failures or successes was computed by dividing the number of mice killed (before the kittens were 56 days old) from the first kill on (Table 1, column 10) by the number of mice presented (column 9). These "killing rates" are given in column 11; they can better be first discussed with the help of Table 4, in which these quotients have been classified in 11 categories.

Table 4. Number of mice killed/number of mice presented, with and without mother present during test. (Abstracted from column 11 in table 1).

Killing ratio (%)	0	1-10	11-20	21-30	31-40	41-50	51-60	61-70	71-80	81-90	91-100	Nr of kittens in test
Mother present	3	–	–	1	1	1	5	3	5	4	4	27
Mother absent	1	–	1	–	2	1	–	–	3	5	–	13
Total	4	0	1	1	3	2	5	3	8	9	4	40

The table shows that for more than 50% of the kittens the killing rate exceeds 0.70. Of the nine kittens scoring below 0.40, four had the same mother *(cy)* and were tested in her presence. She had the habit of rapidly catching the mouse herself, turning away from the kittens, and eating it (see p. 44). She even sometimes took the mouse from the kittens; her behaviour must have reduced the killing rate of her young. Three other of these nine kittens are the kittens *75, 77* and *79* (with *76* and *78* the only kittens that were tested without littermates present; see Table 1, column 13). It has been mentioned under "methods" that this situation was found unsuitable for these experiments because kittens isolated from their mother and littermates usually show a high tendency to escape and are therefore little interested in prey. This phenomenon could certainly account for the low killing rates in kittens *75, 76, 77* and *79*; why it was absent in 78 we cannot explain.

Table 4 shows no positive effect of the presence of the mother during the tests on the killing ratio. Consistent differences in rank between the kittens were not

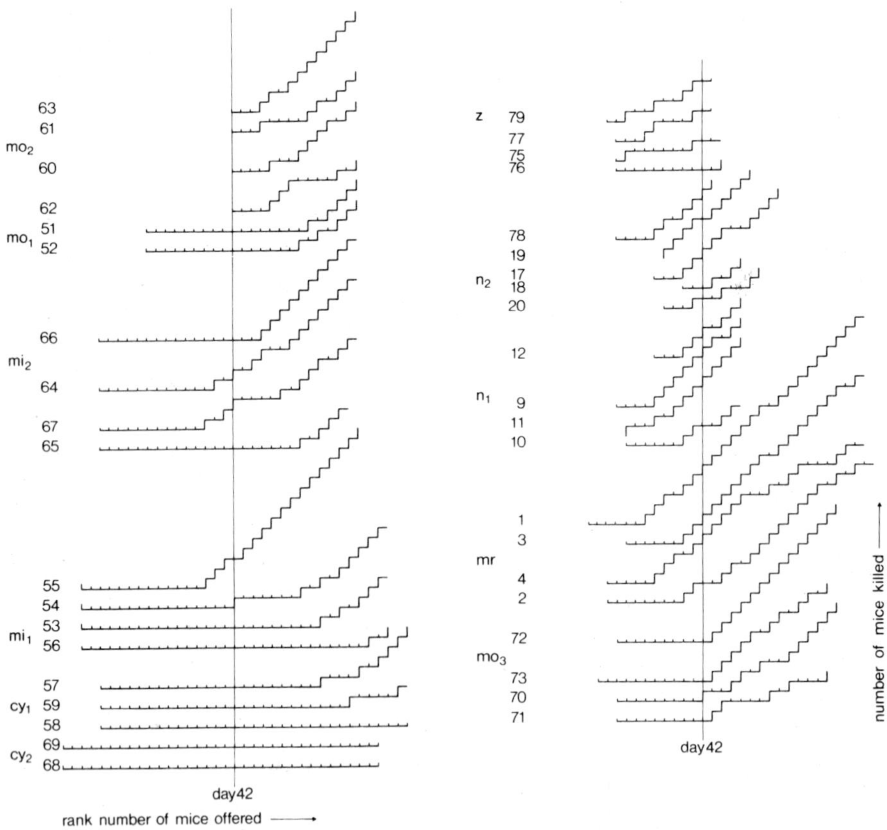

Fig. 9. Cumulative diagrams representing the results of each presentation of a mouse to each of the kittens used in Series I. The individual diagrams are all arranged with reference to day 42; the majority of first killings taking place between day 35 and 49. Each unit on the abcissa means one mouse presented, each unit on the ordinate one mouse killed.

observed during the period the experiment lasted and are therefore unlikely to have influenced the killing ratio. Neither were relations found between the killing ratio and sex (column 14) or condition (column 15), measured as weight increase between the 2nd and 6th week.

The success of each individual kitten over the entire period can be read from column 11 in Table 1. Further for each kitten a cumulative performance diagram is given in Fig. 9, which makes it possible to judge whether in the course of this period changes in performance occurred. Of the 35 cumulative performance diagrams given in this figure those of 12 kittens *(1, 3, 9, 12, 17, 19, 51, 55, 63, 67, 72, 78)* show an almost straight course, which means that no changes took place in the killing ability of these kittens after they had killed their first mouse. In 4 cases *(2, 4, 61, 73)* the performance tended to recede during the period of the experiment and in 3 cases *(20, 59, 71)* it was low from the beginning. In 11 cases *(11, 52, 53, 54, 57, 60, 61, 62, 64, 67* and *70)* the performance improved in the course of time. In 6 cases *(10, 18, 53, 56, 59, 65)* the period during which mice were killed in tests were too short to make any conclusion possible. Litter *z* has been left out of consideration here because these kittens were tested singly. The data of the experiment so far discussed give no indications on possible learning processes before the first mouse is killed.

It has often been suggested that the mother might play a role in learning processes concerning prey. Table 2 shows that the presence of the mother during confrontation with the mouse (see column 12) is not necessary for the development of prey catching and killing behaviour. On the contrary, the litters tested in absence of the mother (the kittens of which had never seen the mother handling a prey) killed their first mouse on the average earlier than those tested when the mother was present; this was independent of the age at which the tests were started.

To understand this apparently inhibitive influence of the mother on killing we have to compare in more detail the behaviour of the kittens in tests with and without mother. The column 6 of Table 1 shows that kittens without mother only very rarely take a mouse in the jaws without sooner or later killing it, whereas this commonly occurs in tests with the mother present. This difference is caused by the interference of the mother when the kitten does not manage to quickly kill its prey. In contrast to the mother, littermates in general do not get a chance to interfere because the kitten that has captured the mouse in the jaws keeps the littermates off by growling and slamming the frontlegs with outstretched fingers and nails.

Having killed the prey the mother as a rule gives the kittens the opportunity to eat from it (see p. 36) but rarely does a kitten get an opportunity to eat from a prey caught by a littermate (Table 1, column 7). After the kittens are 6 weeks old the mother has much less chance to interfere; on the contrary if she happens to hold the mouse the kitten forces itselve upon her, prompting her to release the mouse before or after she has killed it.

The mothers show individual differences in their response to the mouse and the kittens. In particular between *mr, mo* and *mi* on the one hand, and *cy* on the other these differences were marked. When the former three cats caught mouse among the kittens, they did so hesitatingly and easily released it again. In contrast *cy* always rapidly snatched the mouse, turned herself away from the kittens and ate the prey. The *cy*-kittens even developed the habit of making escape responses – in particular huddling together – whenever a mouse was presented. The differences between the four mothers might be understood on the basis of differences in the experiences they must have had when they were kittens themselves. Mother *cy* was originally kitten mr_1-*4* in the litter mr_1, when she grew up in the laboratory where she obtained a very effective routine in catching in the experimental situation. The mothers *mi* and *mr* grew up at home, caught their mice in the wild and had no previous experience with mice in the laboratory. Mother *mo* grew up in the laboratory, but in isolation, and never killed a mouse before she had her first litter (mr_2-*13* in Table 7).

Littermates though, often disturb one another when catching mice, for instance by being in the way or by distracting attention. On the other hand the above mentioned aggression, aroused by littermates in a kitten that has got hold of a prey, may be redirected towards the prey and thus facilitate killing.

Looking at columns 6 and 7 of Table 1 we find that in 7 kittens experience in handling mice with the jaws in the period preceding the first kill was obviously unnecessary. Two of these kittens had never seen another cat handling a mouse. Kittens *11* and *19* killed their first mouse at ages 27 and 39 days, respectively, without having ever seen a mouse. Five other kittens did so at ages varying between 33 and 45 days after having seen but never caught or eaten mice *(12, 17, 18, 20, 76).* This means that the motor pattern for catching a mouse with the jaws and biting it to death, is present in a kitten independently of experience with this prey. Does this imply that the possibility of gaining experience during the nesting period is of no importance for catching and killing prey in later life? Such a wide-scoped conclusion would not be justified on the basis of the data presented here because of the special conditions of our experiment, viz.: (1) the kittens were familiar with the test situation and with the experimenter, consequently their exploratory behaviour was not disturbed when the mother was absent; (2) only one species of prey was used; (3) the mice presented were of a size which the kittens could manage; (4) there was no opportunity in the cage for the mice to escape or hide.

Kittens in an unfamiliar environment in absence of the mother and littermates are frightened and do not handle, catch, or kill prey. The influence of unfamiliar surroundings came out very clearly when we gave a number of kittens (not used before in other experiments and not included in Table 1) without the company of mothers or littermates the opportunity to obtain experience with mice in a much larger space (the ''arena'' used in Series II and described on p. 53) than the cubical cage of 1 m per side they were used to. Of the 9 kittens involved 8 used the 10 minutes of the test entirely for running along the walls, while mewing. Only 1 kitten caught a mouse in the first test and consequently killed all mice in the 5 following tests between the ages of 41 and 55 days. Another kitten killed the first mouse in the 7th test when 53 days old, but only when it was left in the room for a double test period of 20 minutes. In the 2 tests on following days it also killed the mouse. Of the other 7 kittens 2 killed for the first time on the 56th day and again on 2 following days. In the remaining 5 kittens no improvement of catching behaviour was observed in the period they were kept under observation.

Judging from the frequency of the escape responses and the kind of vocalizations uttered the tendency to flee in these kittens appears two have dominated all other behaviour systems. One could think that if this happens repeatedly and for long periods with respect to exploratory behaviour lasting effects on the predatorial behaviour of the cats will ensue. The present experiment was not continued sufficiently long to investigate this possibility.

As to the questions whether during the "nest-period" important experience is gained with the kinds of prey (with regard to species as well as size) suitable for cats, the proceedings described above in cases in which kittens failed to catch a mouse strongly suggests that kittens during the nest-period learn to avoid dangerous prey or to handle them more carefully. In this context we think that the interference of the mother with prey catching during the early period, which appeared to inhibit the killing of the test mouse, may be very functional if the kitten is dealing with a big mouse or a rat. When comparing the behaviour of kittens towards big and small mice, or individual mice which have bitten them before, it becomes clear that the maximum attack behaviour (for which experience with mice is evidently not essential) is as a rule inhibited by escape behaviour, matched to the ferocity of the prey. In short: experience with prey would not be necessary to perform the action pattern in a standard form, but would be necessary to adapt them adequately to the subject they confront. In order to test this hypothesis the experiments of Series II were undertaken, dealing with the effect of deprivation of specific experience during the early period on predatorial behaviour in later life. The treatment of these experiments follows in III.2.2.2.

III.2.2.1.2. *Training on fish as prey (Series IB)*

III.2.2.1.2.1. Methods

The litters n_0 with 4 kittens and mr_3 with 3 kittens were kept with the mother in the laboratory cages and under all other conditions described in the foregoing Series I for the experiments in which kittens were trained on mice. However, this group never received mice as prey. From the 4th up to the 10th week fish were presented to the littermates together, but without the mother; this time not in the cubical cage of 1 m per side, but in a small room familiar to the kittens. Before the kittens were brought into this room four rectangular transparant plastic trays of different sizes had been placed on the floor. The largest tray measured 12 × 20 × 20 (height) cm; it was the standard aquarium vessel also used in all presentations of Series II. The others were, 10, 6 and 4 cm high respectively; the lower they were the smaller was also their surface area, but this remained sufficient for allowing free movement of the head of the kitten within the tray. The water level in the trays reached 2 cm below the edge; in the largest tray water heights of 15 and 12 cm were also used. To make it easy for the kittens to look down into the water and touch it, bricks were placed outside against the higher trays. In most cases cichlid fish (surplus specimens from other behaviour studies) were presented; some roach and carp were used only in the earlier tests of this series. The size of the fish presented varied from 4 to 14 cm (nose to end of tailfin) in correspondence with the size of the growing kittens.

Contrary to Series IA where the initiative was left to the kittens, in this series the procedure during the sessions was aimed at helping the kittens to achieve fish-catching as quickly as possible. This assistance was necessary because of the considerable difficulty the kittens had in overcoming their fear of water. At the start of the session one fish was placed in a tray with a moderately high water column with respect to the size of the kitten. If necessary the attention of the kitten was drawn to the fish by keeping the latter moving. If at least one kitten had paid continuous attention to the fish for 5 minutes without managing to reach it, the fish was transferred to the tray with the next lowest level. If this attempt was also unsuccessful after another 5 minutes attention by at least one of the kittens, the fish was transferred again to a lower water level and this procedure went on until the fish was taken. If necessary the fish was ultimately given on a shallow dry dish. If not scared by the situation the kittens from the beginning were willing and eager to take the fish, fish odour being a powerful attractant for the kittens even at a very young age. With this procedure in each session of Series IB sooner or later every kitten angled a fish successfully (with the exception of kitten *08*).

The mothers were kept away during the training test because whenever they caught a fish they were not willing to share it with the young, in contrast with their behaviour during the training tests for catching mice.

III.2.2.1.2.2. Results

After about 10 weeks 6 out of the 7 kittens involved had developed the same stereotyped pattern, which can be described as follows.

The fish was always taken out of the water by angling with one leg. The nails were extended and placed in the fish by beating it sideways with some force. The fish was then pressed against the vertical wall of the tray and shifted upwards. Only after having brought the fish above the water surface did the kitten begin to use its mouth for gripping the fish in the back, taking care not to wet its nose. The fish reacted to the bite by twisting the body vigorously. After laying the fish on the floor the kittens started eating from the ventral gill area towards the back, during which the movements of the fish gradually waned until finally the spinal cord was cut. Then the kittens mostly first ate the head and finished with the rump, but sometimes it happened the other way round; this choice of direction was not individually different. The fish were always eaten entirely, bones and scales included, unless the portion was too large for the kitten. Evidently the young kitten needs no prior experience with fish to start eating this food when confronted with it outside the water. The problem is how to catch and capture it when the fish is submerged. In the early sessions the kittens first tried to grip the fish with the mouth and thus hit the surface of the water. This sometimes led to a drinking response and always to sneezing and intensive grooming of the nose area, but did not help the kitten to catch the fish. The strong tendency not to wet the nose never waned. Attempts to approach the fish from the side (the transparent wall of the tray) were rare, perhaps because in the early sessions only presentations in the low trays were effective and they may have shaped the habit of approaching the fish from above. After the attempts with the mouth had failed the kitten switched to using a leg, but as soon as the leg got wet it was withdrawn and shaken. However, the kitten was inclined to repeat the attempt and might habituate to the repelling sensation, in particular when through transference of the fish to a shallower tray a following, equally intensive attempt was rewarded. It would gradually accept that an even larger proportion of its leg became wet. The kitten became increasingly motivated to immerse its leg deeper and it developed more skill in chasing and gripping the fish with it. In addition to the stereotyped angling pattern some individual differences were apparent. A large male kitten who, because of sufficiently long legs, had not the slightest difficulties with the height of the vertical walls of the largest tray, caught its fish with striking ease. In contrast a small, but in several aspects very courageous female, jumped into the aquarium where it stood on its four legs, the water reaching its belly (but not wetting its nose). Once inside the tray she followed the same procedure of shifting the fish upwards with one leg, just as the other kittens.

Table 5. Data from the tests with litter n_o in Series IB.

Kitten nr.	05	06	07	08
Age on which first fish was presented; days	31st	31st	32nd	32nd
Number of fish presented before first successful angling	14	30	14	25
Number of fish eaten before first successful angling	11	26	12	21
Age at first successfull angling; days	49th	49th	49th	106th
Number of fish presented between first successful angling and day 56	6	6	4	0
Number of fish angled in same period	7	7	5	0
Total number of fish eaten in same period	18	33	17	12

Some quantitative data for the development of fish catching are given in Table 5 for the litter n_0. It is striking that the first successful angling of the three kittens achieving this performance within 56 days, took place when they were 49 days old. This stands in contrast to the considerable variation in the age at which different kittens killed their first mouse (see Table 1, column 4). They managed to catch the fish in 8 cm deep water. After this first successful attempt they never failed again, although they were sometimes not immediately successful at the highest water level presented. Because the training of litter mr_3 had to be done by a temporary assistant no quantitative data comparable with those of litter n_o are available. However, the 3 kittens were all able to catch fish when the water depth was 8 cm at the 65th day. Then kitten *14* had been offered 10 fish, kitten *15* got 17 fish and kitten *16* had 11 opportunities. All fish presented were eaten. Also in this series the kittens took to angling around the 50th day. At that age an improvement takes place in the co-ordination between the mouth and the leg, facilitating the passing of the fish from the latter to the former.

The tests with the litters n_o and mr_3 were continued after the 9th week, as a part (exp. Q) of Series II. Except for kitten *08* all kittens of these litters were then able to catch any fish offered in 8-12 cm deep water. Kitten *08* sometimes put a leg into the water but never deep enough. It regularly got fish, though partly by stealing it from its mates. However, when 106 days old it first succeeded in catching a roach from an aquarium with 8 cm water. On the 109th day it angled a roach from a depth of 15 cm and from then on was as successful as the others at every further presentation. Judging from its behaviour *08* seemed more afraid than the other kittens of getting a wet leg and managed to compensate this disadvantage by developing successfully the habit of stealing fish from littermates.

The stereotyped shape of the angling pattern in spite of differences in size between the kittens, and in contrast to the acquisition of different additional habits such as jumping into the tray, argues in favour of the idea that the motor pattern angling as such has not to be acquired from experience with fish (see p. 36). Also the fact that *08* finally performed the pattern in the same form, but only adopted it at a much older age than other kittens, supports this view. Moreover, Leyhausen (1965[a]) has described angling in cats capturing mice from narrow crevices. We also occasionally observed this in the "home and garden" situation; in our experimental situation there were no crevices available for the mice to hide. Nevertheless, a considerable amount of learning seems to be possible in the manner and situation in which the pattern is used.

For the identification of fish as suitable food no experience is necessary. However, when kittens familiar with roach or carp were given cichlids (or the reverse) they tended to play around with the prey for some time after having captured and laid it on the floor. We think that this was caused by some inhibition to start eating due to unfamiliarity with that kind of food.

III.2.2.2. Series II: The later period

III.2.2.2.1. *Design*

These experiments were carried out with kittens of at least 8 weeks. At 56 days, the mothers had been removed from the home cages, whereas the littermates remained together in these cages which were connected with an outside pen as described in I.2.b. Series II consists of seven experiments (M, N, P, R, S, and Q and T); a survey of them is given in Table 6.

Table 6. Survey of the experiments in Series II.

Code of litter and kittens living with nursing mother until weaning	Exp. Code	Experience during "early period" < 56 days		Tests in "later periods" (over 8 weeks)		
		prey species	prey numbers	age in days	prey given	
					mice	fish
mr_1: 1, 2, 3, 4.	M, Q	mice	30-34	56-277	72-97	19-45
n_1: 9, 10, 11, 12.	N	mice	10-14	73-238	18-26	none
n_2: 17, 18, 19, 20.	N, Q	mice	7-14	57-169	21-24	15-20
n_0: 5, 6, 7, 8.	P, Q	fish	18-36	84-280	15-36	33-42
mr_3: 14, 15, 16.	P, Q	fish	10-17	65-195	12-13	43-51
mr_2: 13.	R	none	–	120-211	13	none
br: 21.	R, Q	none	–	102-169	12	7
sg: 27, 29, 30, 31.	S	none	–	125-219	24-26	none
mi_2: 33, 34, 35, 36.	S	none	–	98-185	26-28	none
cy_0: 37, 38, 39.	S	none	–	84-159	25-26	none
mi_1: 22, 23, 24, 25.	T, Q	experience in garden	?	101-168	11-13	17-21

During the early period exclusively mice of 15-25 g were presented to the kittens in the experiments M and N (three litters, each of four kittens), as described under methods in Serie IA. Presentation occurred with the full litter present. Under the age of 56 days in exp. M the prey was offered in the presence of the mother, and in both nests of exp. N in her absence (Table 1, litters mr_1, n_1 and n_2).

In exp. P – involving two litters of 4 and 3 – the kittens obtained during the early period exclusively live fish in small aquaria filled with water (for methods used in training see III.2.2.1.2).

In exps. R and S no prey had been given to the kittens until the tests of Series II started. Exp. S was carried out with three litters, containing 4, 4, and 3 kittens, respectively. In exp. R two litters were used, in each of which only one kitten was present. During the first 7 weeks the nursing mother had been left with the kittens; thereafter she was taken away and the kitten stayed isolated for the rest of the period in a laboratory cage without outdoor pen.

The effect of the differential treatment given this way was in all groups tested by presenting a mouse to each kitten individually on several days spread between the 2nd and 9th month (as recorded in Table 6).

The 9th month was chosen as an upper limit because after about 280 days of age the kittens began to show sexual behaviour. The interference of the sexual system introduced a new variable and made the results more difficult to interpret and less comparable with those obtained before that time. For instance, with the increase of the sexual factor the eagerness to eat went down and the cats became more gentle not only towards their littermates, but also towards the mice, rubbing the mice with their heads and rolling with them on the floor, as they would with littermates.

In exp. Q a number of kittens used in the exps. M, N, P and R were after day 100 and during the period they were regularly confronted with mice, also from time to time presented with fish (see Table 8). In these tests the fish were usually given in the aquarium and then always with the water at the maximum level. However, to encourage the kittens dead or live fish were sometimes also offered outside the tray on the dry ground (see p. 58).

In exp. T – involving one litter *(mi$_l$)* of four kittens – the mother and kittens were kept at home with access to the garden up to the age of 14 weeks and then transferred without the mother to the laboratory situation. Therefore, the kittens of this group can be taken to have had experience with a variety of stimuli from the natural environment, of which stimuli from butterflies, bumblebees, beetles, grashoppers, frogs, newts, slugs and mice (the latter mostly caught and brought in by the mother) are of particular interest in this context (see p. 36). In a way exp. T serves as a control. In the laboratory flies and crickets were the only animals towards which hunting behaviour might occur. Flies were never, but crickets sometimes, caught and eaten. To the kittens of exp. T mice and fish in the aquarium filled with water were offered in the same way as described above in the experiments with the other kittens of Series II (see Table 6, 7 and 8).

III.2.2.2.2. *Methods of testing*

As described in III.2.2.1. for Series IA all tests were done between 8.00 and 14.00 hour. The kittens were fed 16 hours before and shortly after the tests. Around 8.00 in the morning before they were tested milk was provided.

The only difference was that the kittens were tested one by one and in an ovally shaped room 8 × 4 m with a diffuse illumination (the arena). A kitten entered the room through a sliding door, opened from a distance by the experimenter who was sitting outside the arena behind an observation window.

Some days before the experiments started the kittens were allowed for 30 minutes to play in this room with their littermates. When they entered the arena to be tested they did not show signs of fear. After some experience with the tests they showed great eagerness to enter, rushed in, and after having finished they rushed out again at the first sound of opening the sliding door to be brought back to their homecage.

Before a kitten was allowed to enter a mouse was put into a rectangular plastic aquarium (12 × 20 × 20 cm) placed in the center of the arena. In the exps. M, N and P the open side of the tray was on top, in conformity with the tests of exp. Q in which the cats obtained fish in the same sort of tray filled with water. To make sure that their lack of experience with this tray would not prevent the kittens of the litters R, S and T from coming into contact with the mouse, in these experiments the tray was placed on its side. As a consequence in these tests the mouse was only for a while kept in a fixed place; it could leave it and run around freely in the arena.

In the protocol of each test the time was recorded at the following moments:

— the introduction of the prey into the test cage (t_o)

— the first grip on the prey with the mouth irrespective of whether it was later on released (t_1)

— the killing bite (t_2)

— the start of eating (t_3)

— the end of eating, irrespective of whether the cat had completely finished the prey or not (t_4).

The intervals between these instances are called p_1, p_2, p_3 and p_4, respectively, for the entire test period the symbol p is used. The criterium for the choice of these periods was that each of them was different with respect to the stimuli emanating from the external situation and perceivable to the kitten. During p_1, as long as the prey was sitting, walking, or running the kitten could receive stimuli from the moving object which also might produce sounds. The kittens reacted by following the prey and touching it with the paw in various ways. During p_2 the kittens sniffed at the prey and nibbled at it with the mouth. If the prey was taken into the mouth and lifted from the floor (= caught), its wriggling movements could be perceived by the lips and in particular by the whiskers. In addition at this stage taste reception became possible. If the prey managed to escape in p_2 a repetition of the behaviour of p_1 mostly occurred. However, if the bite was strong enough to kill the prey, period p_3 had started. The handling of the dead prey now became easier. The immobility of the prey is an important change in this situation. In addition blood might flow and provide new smell and taste stimuli to the kitten, who often licked it up from the prey as well as the floor. In p_4 the skin is torn, after which the prey is devoured piece by piece.

Catching, killing and eating may follow one another without interruption by other behaviour. Often, however, this sequence was interrupted by behaviour not directly subserving these predatory functions.

A shorthand account of the occurrence of such activities during each of the intervals was made. In working up these records attention was particularly paid to frequency per test of each of the following behaviour patterns:

— growling

— hissing

— other vocalizations (including mew and prr)

— walking without prey

— grooming (of various kinds, scratching, licking, nibbling)

— carrying the prey around

— touching the prey with the forepaw

Table 7. Survey of results of Series II, specified per experiment and per kitten

Animal and age	Total nr tests	Caught		Killed		Start eating			Finished eating	
		total nr	time <1 min	total nr	time <3 min	total nr	time <2 min	time <5 min	total nr	time <2 min
EXPERIMENT M										
mr_1 1 56-277 days	97	97 100%	83 86%	95 97%	88 93%	95 97%	87 91%	90 95%	93 96%	76 81%
mr_1 2 56-277 days	78	78 100%	65 83%	71 91%	49 69%	71 91%	71 100%	71 100%	71 91%	48 68%
mr_1 3 56-277 days	72	71 98%	64 90%	66 91%	36 54%	66 91%	64 97%	66 100%	66 91%	43 65%
mr_1 4 56-277 days	68	67 98%	47 70%	47 70%	19 40%	46 68%	44 96%	44 96%	45 66%	25 56%
Exp. M total in %	315	313 99%	259 83%	279 89%	192 69%	278 89%	266 96%	271 97%	275 87%	192 70%
EXPERIMENT N										
n_1 9 73-238 days	23	23 100%	20 87%	22 96%	10 46%	22 96%	20 91%	22 100%	22 96%	14 63%
n_1 10 73-238 days	26	26 100%	25 87%	24 93%	10 42%	24 92%	14 58%	18 75%	24 92%	15 63%
n_1 11 73-238 days	18	17 94%	15 88%	17 94%	7 42%	17 94%	13 77%	15 88%	17 94%	7 42%
n_1 12 73-238 days	25	24 96%	21 87%	24 96%	14 58%	24 96%	18 75%	20 83%	24 96%	15 62%
n_2 17 57-169 days	21	20 95%	19 95%	19 92%	18 94%	19 92%	19 100%	19 100%	19 92%	4 21%
n_2 18 57-169 days	21	21 100%	19 91%	19 91%	17 90%	18 86%	17 94%	17 94%	18 86%	1 5%
n_2 19 57-169 days	24	23 96%	20 87%	19 80%	17 90%	19 80%	19 100%	19 100%	19 80%	2 10%
n_2 20 57-169 days	21	20 95%	15 75%	20 95%	19 95%	20 95%	19 95%	20 100%	20 95%	3 15%
Exp. N total in %	179	175 98%	154 88%	164 92%	112 68%	163 92%	139 85%	150 92%	163 92%	61 43%
EXPERIMENT P										
n_0 5 137-277 days	15	12 80%	11 91%	0 0%	0 0%	0 0%	0 0%	0 0%	0 0%	0 0%
n_0 6 137-277 days	21	19 90%	18 95%	15 72%	11 73%	13 62%	11 84%	13 100%	13 62%	11 84%
n_0 7 137-277 days	23	17 74%	11 62%	14 61%	13 93%	4 18%	4 100%	4 100%	4 18%	2 5%
n_0 8 137-277 days	20	11 55%	8 73%	8 40%	6 75%	3 15%	1 33%	2 66%	3 15%	0 0%
mr_3 14 125-195 days	12	12 100%	10 84%	11 87%	5 45%	11 87%	9 82%	11 100%	11 87%	1 9%
mr_3 15 125-195 days	12	12 100%	12 100%	12 100%	10 84%	9 75%	7 78%	7 78%	9 75%	1 11%
mr_3 16 125-195 days	13	13 100%	11 84%	11 85%	10 92%	9 69%	8 88%	9 100%	9 69%	4 45%
Exp. P total in %	116	96 83%	81 84%	71 61%	55 77%	49 42%	39 80%	46 93%	49 42%	19 39%

Table 7. Continued.

Animal and age	Total nr tests	Caught total nr	Caught time <1 min	Killed total nr	Killed time <3 min	Start eating total nr	Start eating time <2 min	Start eating time <5 min	Finished eating total nr	Finished eating time <2 min
EXPERIMENT R										
mr_2 13 120-211 days	13	0 0%		0 0%		0 0%			0 0%	
br 21 102-169 days	12	1 8%	1 100%	0 0%		0 0%			0 0%	
Exp. R total in %	25	1 4%	100%							
EXPERIMENT S										
sg 27 114-233 days	25	19 76%	17 89%	18 72%	16 88%	15 60%	8 53%	9 60%	15 60%	3 20%
sg 29 125-219 days	25	0 0%	0 0%	0 0%	0 0%	0 0%	0 0%	0 0%	0 0%	0 0%
sg 30 125-219 days	26	17 65%	15 88%	16 62%	15 93%	14 54%	4 29%	6 43%	14 54%	2 14%
sg 31 125-219 days	24	21 82%	18 86%	16 67%	10 62%	14 58%	11 78%	13 93%	14 58%	2 14%
mi_2 33 98-185 days	26	11 42%	10 91%	8 31%	3 38%	7 27%	2 28%	4 57%	7 27%	1 14%
mi_2 34 98-185 days	28	23 82%	22 96%	23 82%	19 83%	23 82%	23 100%	23 100%	23 82%	7 31%
mi_2 35 98-185 days	28	1 3%	0 0%	0 0%	0 0%	0 0%	0 0%	0 0%	0 0%	0 0%
mi_2 36 98-185 days	27	5 19%	0 0%	0 0%	0 0%	0 0%	0 0%	0 0%	0 0%	0 0%
cy_0 37 84-159 days	25	10 40%	5 50%	0 0%	0 0%	0 0%	0 0%	0 0%	0 0%	0 0%
cy_0 38 84-159 days	26	26 100%	24 92%	26 100%	24 92%	26 100%	21 81%	23 88%	26 100%	2 8%
cy_0 39 84-159 days	26	1 4%	0 0%	0 0%	0 0%	0 0%	0 0%	0 0%	0 0%	0 0%
Exp. S total in %	286	134 47%	111 83%	107 37%	87 82%	99 35%	69 69%	78 78%	99 35%	17 17%
EXPERIMENT T										
mi_1 22 101-168 days	11	4 36%	2 50%	4 36%	3 75%	3 27%	3 100%	3 100%	3 27%	0 0%
mi_1 23 101-168 days	13	11 84%	3 27%	5 39%	3 60%	3 23%	0 0%	1 33%	3 23%	0 0%
mi_1 24 101-168 days	13	11 84%	7 64%	5 39%	4 80%	5 39%	3 60%	3 60%	5 39%	1 20%
mi_1 25 101-168 days	13	6 46%	4 66%	4 31%	0 0%	4 31%	3 75%	4 100%	4 31%	0 0%
Exp. T total in %	50	32 64%	16 50%	18 36%	9 50%	15 30%	9 60%	11 73%	15 30%	1 7%

Table 8. Survey of the results of experiment Q in Series II, specified per experiment and per kitten.

Litter nr.	Kitten nr.	Presented before day 100	Eaten (dead or alive) before day 100	Angled from water before day 100	Presented day 100-280	Eaten (dead or alive) days 100-280	Angled from water days 100-280	Presented days 280-315	Eaten (dead or alive) days 280-315	Angled from water days 280-315
mr_1	1	0	–	–	19	1	0	12	4	0
	2	0	–	–	45	8	6	11	7	10
	3	0	–	–	22	3	0	13	6	0
	4	0	–	–	43	0	0	14	5	0
n_2	17	0	–	–	19	5	0			
	18	0	–	–	17	3	0			
	19	0	–	–	15	4	0			
	20	0	–	–	20	2	0			
br	21	0	–	–	7	0	0	discontinued at day 176		
mi_1	22	0	–	–	17	2	0			
	23	0	–	–	17	0	0			
	24	0	–	–	21	0	0			
	25	0	–	–	21	0	0			
n_0	5	39	26	20	32	27	27	3	3	3
	6	52	44	19	34	33	32	1	1	1
	7	35	30	11	25	24	24	1	1	1
	8	22	19	0	29	26	23	4	3	3
mr_3	14	42	39	25	19	19	17			
	15	44	40	24	20	20	20	discontinued at day 200		
	16	37	34	23	17	17	17			

tossing the prey (with the mouth)
head rubbing (against the floor, the walls of the cage or the prey)
rolling (see for description p. 64).

On the basis of the above records the performance of each particular kitten was expressed by a number of different parameters, viz.:

— total number of mice caught with the mouth within the ten-minute period of all tests and number of these mice caught within 1 min.;

— total number of mice killed in the test period and the number killed within 3 min. after catching with the mouth;

— total number of times a kitten started eating the mouse; with in addition the number of times eating started within 2 min. and within 5 min. from the moment the kitten killed the prey;

— total number of times a prey was entirely finished within the entire test period and within 2 min. after eating started. To make the results comparable all parameters are also expressed as percentages of the maximum values theoretically possible.

The interruptive behaviour patterns will be discussed in III.2.2.3.

III.2.2.2.3. *Results*

For the exps. M, N, P, R, S and T the data for each individual kitten are given in Table 7. An overall survey of these results is given in the histograms of Fig. 10. The data for exp. Q can be found in Table 8.

Between the results of the exps. M and N no appreciable differences were found, neither in the percentage of occasions mice were caught, killed and eaten (Fig. 10a), nor in the speed at which these activities were carried out (Fig. 10b). All parameters reached very high values; the individual differences were not large. The relatively low mean percentage and the large spread for the tests in which eating was finished within 2 min. is due to the fact that the four kittens in the n_2 litter were only tested when less than 170 days old, while all mr_1 kittens

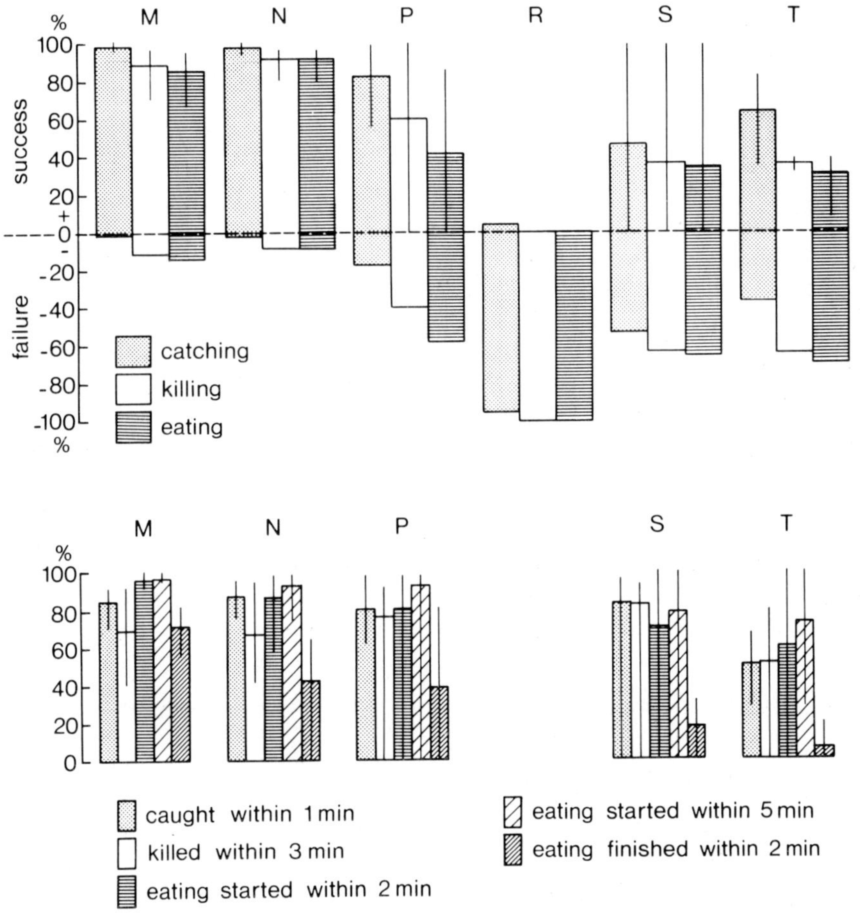

Fig. 10. Representation of the overall results of six experiments of Series II.

a) in terms of the percentage of presentations of mice in which the prey was only caught, caught and killed, or caught, killed and eaten.

b. in terms of the promptness with which catching, killing and eating were carried out, represented as percentages of all presentations in which these activities took place within the time limits indicated. The vertical lines in each column indicate the total spreading of the percentages per individual kitten in the experiment concerned.

were tested until day 277, that is when they were also much stronger. We may conclude that all cats which had had experience with catching mice in Series IA were successful mouse-killers in Series II. Presence or absence of the mother during the training tests of Series IA made no difference in tests of Series II. The correspondence between the results of exps. M and N made us decide not to maintain the high number of mouse-presentations chosen for exp. M in the other experiments.

The two isolated kittens in exp. R were both unsuccessful; they never caught a mouse with the mouth, except for *br-21*, for a short instance in the last of 12 tests. During most tests of this experiment the mice were touched with a frontleg, which was stretched out while the head was withdrawn. This touching became more vigorous in the course of the experiment when running (chasing the mouse) also increased, both in frequency and in intensity. There were, however, interesting individual differences between both kittens; *br-21* showed more and very violent chasing, mr_2*-13* more touching (see also p. 63 and 80 and Table 14).

In the exps. S and T the overall performance was considerably lower than in the exps. M and N (Fig. 10a). The activities of the kittens in these experiments can be compared in more detail by means of the data in Table 7 and the histograms of Fig. 10b. The figures show that relatively low performance is due to two phenomena: a delay until the first mouse is killed and a strong individual variation in this delay (from occasional observations of the behaviour of some of the non-successful kittens in later life we have reasons to believe that all kittens would succeed if the number of tests and/or the period would be extended). It is striking that from the first time a mouse is killed and eaten the kitten has become an efficient mouse-killer and hardly ever fails again.

It can be seen in Fig. 10b that the average speed at which the kittens in exp. S caught, killed and started to eat the mouse were not different from those of kittens in exps. M and N. The S-kittens, however, were much slower eaters. This can partly, but not wholly be attributed to their lower average age. If kittens of the same age are compared between exps. S and N the average time for eating a mouse was 6.6 and 4.2 min, respectively; this difference was chiefly due to longer eating times of the S-kittens during the first 10 tests. In contrast to the M- and N-kittens the S-kittens had never been given mice before.

The most striking difference between exps. S, M and N is the variation in performance between the individual kittens, which was large in S and small in M and N. The individual variation in performance between the kittens in exp.S cannot be attributed to the age at which the first mouse was presented. Kittens completely failing to catch occurred in tests started at the age of 84, 98 and 125 days. Differences in social rank between littermates may be involved, but we have not succeeded in establishing such differences and correlating them with individual performances; if they exist they are certainly not simple and evident. Differences in "personality" which may be relevant in this context do exist between kittens. Anyway, the experience obtained during the training in exps. M and N reduces the variation dramatically.

The observations suggest that only one opportunity to eat from a mouse may suffice to make a kitten a successful mouse-killer. In exp. S this opportunity was probably randomly distributed over time, in the exps. M and N it was greatly increased because of the training procedure.

In exp. T none of the kittens killed and ate a mouse during the first seven presentations of a living mouse, although they sometimes succeeded in capturing it. At the 8th presentation, however, when the 10 min. test period had ended kitten *25* caught a mouse and when the experimenter tried to remove the mouse to discontinue the session this stimulated the killing bite and the kitten was allowed to eat. After this incident kitten *25* had become a successful mouse-killer. Also at the 8th presentation, kitten *22* had captured a mouse, but instead of biting it to death, carried it around awkwardly for some time, as a result of which the mouse died. The kitten, however, did not try to eat this mouse until a second kitten was introduced into the cage. At the 9th presentation this kitten and kittens *23* and *24* obtained a dead mouse. All ate it and from then on apperared to have turned into successful mouse-killers.

As shown in Fig. 10b the T-kittens were on the average slower in catching and killing mice than the kittens in the other experiments. We believe that the relatively long time passing in exp. T before a mouse was captured, killed, and eating had started was a consequence of the way they were brought up, namely with access to the garden. There they became conditioned to approaching a wild living prey cautiously while making use of cover. Correspondingly, in the tests the T-kittens approached a potential prey far more sneakingly than the kittens of the other experiments. Before they killed a mouse, and again before they started eating, they had the habit of walking with the prey for some time before stopping in a place which was apparently satisfactory enough to proceed further with the activity chain of capturing, killing, and devouring the prey.

The kittens of exp. P had been trained on fish only when we started to test them on mice. Their catching performance on mice was just a little below that in exps. M and N (see Fig. 10a). Six of the seven kittens in this experiment caught the mouse at the first presentation. If killing took place the latency was not different from that of the kittens in exps. M, N and S (Fig. 10b), but killing occurred much less frequently (Fig. 10a), often a P-kitten let a mouse escape. Of the mice caught 68% were not eaten, but if eating took place the latency was similar to that in exps. M and N, and the average time spent eating was not significantly longer. Eating a mouse once did not, like in exps. S and T, turn the kittens into quick killers. We are inclined to believe that the fact that in the training tests fish which had been caught could not escape, was a cause for the initially clumsy behaviour of P-kittens which had caught a mouse. In contrast to eating a mouse, eating a fish began without killing it first. Moreover, whereas experienced cats always captured mice first with the mouth, fish were always first angled with a paw. The P-kittens also practiced this technique for taking a mouse from the tray, but this was often insufficient to prevent the mouse from running away. Kittens *14, 15* and *16* of litter mr_3 developed earlier into successful cathers than kittens *5, 6, 7* and *8* of litter n_0, although their behaviour towards the fish was not essentially different. We cannot explain this difference. However, a possibility exists that it is due to slight differences during the training period because, due to illness, the training of litter n_o had to be done by another person.

The results of exp. Q – the presentation of fish to mouse catchers with differently programmed experience (M, N, R and T) – are given in Table 8.

Only one of the 13 kittens involved became a fish catcher. None of the others ever caught a fish in the water in spite of the fact that 8 or them took the opportunity to eat fish when it was presented to them dead or alive outside the water. The successful fish-catcher mr_l*-2* and kitten mr_l*-4* were together in an experiment – that will be more extensively treated on p. 92 – in which they had the opportunity to watch the behaviour of an experienced fish-catcher from exp. P. The example of its mate certainly did not turn mr_l*-4* into a fish-catcher. Whether it had influence on the development of fish catching in mr_l*-2* is difficult to decide. This kitten did not catch its first fish from the water at the first opportunity after having been confronted with the example. It angled successfully only after another 8 fish had been presented; in the following 18 tests, however, it only failed thrice. At a later age the habit was found to have remained, just as in the P-kittens.

From the fact that dead fish or living fish outside the water are readily taken and eaten by most kittens one may conclude that the crucial problem which has to be solved in these tests is how to cope with the water barrier.

III.2.2.2.4. *Conclusions*

Experience in catching and handling prey during the early period is important later for facilitating prey catching, killing and eating, but it is not absolutely necessary for the development of these capabilities. Experience during this nest period also reduces the individual variation between kittens in these skills.

Kittens are able to become mouse-killers without having been trained on these prey during the early period; most socially raised kittens without early experience with these prey which were later exposed to mice were ultimately able to kill them. Only one mouse-catching kitten, however, later developed the habit of fish-catching. A marked individual variation existed between kittens in the number of presentations attended before the habit was acquired. After the first successful killing and eating, however, failures rarely occurred. When this killing happened largely depended on when a suitable opportunity occurred in the tests of Series II. The moment this took place is probably randomly distributed over the different presentations, but is in the training tests of Series I promoted by the procedure, and in kittens growing up under normal conditions by the activities of the mother and of littermates.

The failure of the R-kitttens to catch mice shows that the social experience with the mother and with littermates is also a factor promoting prey-catching. To understand this a discussion of details in the behaviour during the different experiments given in the next section will be necessary.

The switch from fish-catching to mouse-catching is much less difficult than the reverse. Without the training in which the water levels were changed stepwise it is difficult for a kitten to overcome the water barrier. The habits developed after experience with one kind of prey are also used vis à vis the other. In the final technique specific pecularities of the prey, such as attempts to escape or counter-attack, are taken into acount.

III.2.2.3. The occurrence of interruptive behaviour during the tests

The activities dealt with so far play a direct role in catching, killing and eating the prey. During the tests other activities may also occur which to the observer appear as little or non-functional for achieving the consummatory act of he prey-catching sequence. In the context of the incubation behaviour of birds Baerends et al. (1970) have proposed the name "interruptive behaviour" for such activities. They felt the need for a new and neutral term because all terms which in ethology are commonly used for denoting similar phenomena have also obtained certain causal and/or functional meanings, which unfortunately are not used with much consistency by different authors. In this paper we have adopted the usage of the term "interruptive behaviour" for the functionally non-obligatory behaviour patterns occurring during catching, killing or eating a prey. For at least some of them an attempt to explain their causation and/or function seems valuable to us. The following behaviour patterns have been included in this category: growling, hissing, mewing (and other vocalizations), walking (without prey), grooming, carrying, touching, tossing, head rubbing, rolling.

A rough survey of the occurrence of interruptive activities in the various tests can be obtained from Table 9, giving for each of these activities the number of tests in which they were observed (irrespective of the frequency of their occurrence in these tests). The data are not only given for all 1960 tests together but also separately for different categories of tests, viz.: tests with kittens below and above 56 days (which moreover were carried out in a different environment, see p. 38, and 53, and with respect to the younger group, in the presence and absence of the mother). The data have also been subdivided for the four periods distinguished: the time until catching (p_1), the time until killing (p_2), the time until eating (p_3) and the eating period (p_4).

A comparison of the data on interruptive behaviour in tests with different results with respect to catching, killing and eating, would be facilitated if the number of tests in which the various interruptive activities occurred could be expressed as percentages of the total number of tests in the category concerned. However, it was felt that such a procedure is not permissible for most of the experiments, mainly on the ground that sometimes the experimenter had, for various reasons, found it desirable to extend a test beyond the standard 10 minute period. As a consequence of this more opportunity for interruptive behaviour was available in such cases which reduced the comparability of the results between tests. Another difficulty is the varying length of the p-periods. Actually they were determined by the behaviour of the kittens; a kitten passing quickly through a phase had less time for performing interruptive behaviour in that phase than a kitten that hesitated.

Table 9. The occurrence of interruptive activities in the tests of Series I and II. The results of column a are specified with regard to age in the columns b and c and column c is further specified in the columns d and e with regard to presence or absence of the mother during the tests.

Periods in tests	p1					p2					p3					p4				
	a	b	c	d	e	a	b	c	d	e	a	b	c	d	e	a	b	c	d	e
				Before day 56					Before day 56					Before day 56					Before day 56	
	Total	Age > 56 days	Age < 56 days	With mother	Without mother	Total	Age > 56 days	Age < 56 days	With mother	Without mother	Total	Age > 56 days	Age < 56 days	With mother	Without mother	Total	Age > 56 days	Age < 56 days	With mother	Without mother
Number of presentations	1960	1076	884	783	101	1316	799	517	451	66	1192	685	507	454	53	1161	667	494	443	51
Growling	43	21	22	18	4	163	97	66	51	15	78	42	36	23	13	83	38	45	31	14
Mewing	155	107	48	44	4	51	41	10	10	0	20	19	1	1	0	7	4	3	1	2
Hissing	6	3	3	2	1	9	5	4	4	0	5	2	3	3	0	5	1	4	4	0
Walking without prey	345	270	75	65	10	160	136	24	21	3	12	8	4	2	2	3	2	1	1	0
Grooming	134	119	15	15	0	21	18	3	3	0	5	5	0	x	x	0	x	x	x	x
Carrying the prey	x	x	x	x	x	264	212	52	42	10	75	53	22	21	1	27	11	16	11	5
Touching	322	185	137	129	8	313	253	60	58	2	95	67	28	27	1	18	12	6	5	1
Tossing	4	3	1	1	0	115	96	19	15	4	74	53	21	21	0	16	14	2	2	0
Rubbing	57	54	3	3	0	35	33	2	2	0	16	12	4	3	1	1	0	1	1	0
Rolling	65	53	12	12	0	99	96	3	3	0	34	34	0	x	x	2	2	0	x	x

In exp. S, however, the length of the 289 tests never exceeded 10 min. In addition a number of other factors made these tests for our purpose more comparable than those of the other experiments. First, all kittens were over 56 days old, which means that they were physically sufficiently well developed for catching mice. Second, because they were less than 280 days old, interfering influences of sexual factors during these tests are unlikely. Finally, in these tests the kittens were always tested singly so that a possible inducement of interruptive behaviour by littermates was excluded. The data on interruptive behaviour in this exp. S are therefore separately given in Table 10 in absolute numbers as well as in percentages. The different p-periods are not considered separately, but the total time either until the end of the 10 min period, or unil the end of eating is taken. One still can object that because the test was shorter when the mouse was eaten in such tests, less time for interruptive behaviour was avialable. However, it can also be argued that it was nevertheless the kitten who made the decision how to spent its time and that consequently a comparison does make sense, provided it is done with caution. In our opinion with these reservations for exp. S the percentages are useful for a rough comparison of the data, but one should not attribute too much importance to the exact values and to smaller differences between them.

Table 10. The incidence of interruptive activities in tests of exp. S.

Type of "interruptive" activity"	Number of tests with:			
	Mouse not caught		Mouse caught	
	Numbers	%	Numbers	%
Growling	0	*0*	22	*14*
Mewing	79	*52*	23	*17*
Walking	137	*90*	29	*21*
Grooming	94	*62*	13	*9*
Sitting	93	*61*	0	*0*
Carrying	0	*0*	62	*45*
Touching	85	*56*	88	*64*
Tossing	3	*2*	44	*32*
Rubbing	19	*13*	13	*9*
Rolling	22	*14*	52	*28*
Total number of tests in group	152		137	

In addition one has to keep in mind that the occurrence of interruptive activities may be inadequately evaluated in Tables 9 and 10, because only their presence or absence in a test has been taken into account, not their frequency of occurrence. This measure has not been used in this survey because of the considerable difficulties to represent it in a comparable way. The frequency data tend to show so much a spread and their proper quantification is theoretically and technically so difficult that it would ask for a separate study to deal with them appropriately. Although notations about frequency were made this was, as regards exactitude, done with a second priority as compared with the other data. However, on the basis of the data underlying Table 9 and 10, and further quantitative and qualitative observations made during the tests the following comments concerning the causation of the occurrence of the interruptive activities can be made.

Growling. This pattern particularly occurred after the prey was caught, shortly before and during killing. Correspondingly, it was most common in p_2 where it seemed primarily directed against the mouse. The impression from Table 9 that it seemed to occur more often in tests with younger than with older kittens is corroborated by the observation that in the former tests it also occurred considerably more frequently, the young kittens having in general more difficulty in killing the mouse, whereas in the older kittens often one bite sufficed. Growling is more common in young kittens tested without the mother present because the mother tends to interfere when the kitten has trouble with its prey. During training tests with the mother present she also often presented the mouse to one particular kitten, taking it apart in a corner of the cage and thus reducing the amount of interference by nest mates. A considerable amount of growling during p_3 and p_4 is a reaction to the aproach of nest mates (only present in tests under 56 days) when bouts of growling also tend to last for a longer time.

Hissing. This activity may occur in any type of test during any period, but it is not sufficiently frequent for a quantitative treatment. It may be a reaction to a nest mate, but especially in young kittens it was performed when a mouse turned towards the kitten and defended itself.

Mewing. This vocalization was correlated with walking along the walls of the cage, obviously looking for a way to escape from the situation. In young kittens it was frequently connected with a need to urinate, the kitten searching for the tray with shavings or the licking mother, and stopping after the opportunity to urinate had been given. Because it is typical for a kitten not interested in the mouse, it is more common in p_1 than in other periods. It was more common in the category of older kittens because the Series II concerned contained a higher percentage of tests with kittens unfamiliar with mice than Series I and was carried out in less reassuring surroundings (the bigger arena, no nest mates and no mother present).

Walking (without prey). This pattern was only distinguished when walking was more extensive than needed for a straightforward approach of the mouse. The occurrence during p_3 and p_4 is negligible, but it was commonly seen before killing during p_1 and p_2. In these periods it was more frequent in the older than in the younger kittens, although in the latter ones it often took longer before they had spotted a mouse and became interested in it. A considerable amount of walking was caused because the kitten lost its grip on the mouse, let it escape and followed it again. Then walking might turn into fast running. The size of the arena in which the older kittens were tested probably promoted this behaviour, but the occurrence is also connected with the relatively large amount of touching, tossing, head rubbing and rolling in older kittens, which will be discussed below. In the young kittens the presence of the mother tends to reduce this behaviour because of her assistance in preventing the mouse to escape.

Grooming. During prey-catching this behaviour (Pl. VII[b]) was chiefly observed in the older kittens, mainly in p_1 and to a lesser extent in p_2. Of the 142 tests with older kittens in which grooming occurred as a behaviour interrupting prey-catching, 114 tests belonged to exp. S, that is, tests with kittens which had not previously been trained to catch mice. Without showing expressions of fright as strong as retreat or hair raising, these kittens often sat down near the container with the mouse and started licking extensively the flanks, hind legs (in particular the inner side) and the abdomen. If the mouse, after escaping from the container, approached the kitten it might nose at it and follow it with the eye, in a few cases touching it with the paw, but the grooming kitten did not pursue it.

Grooming always occurred very extensively after eating a mouse. Then much attention was given to the head and forepaws. *This* grooming is not incorporated in Table 9 and 10. Long bouts of grooming occurred in cats trained in catching mice who, when offered a fish in the aquarium, had got their nose wet in an attempt to approach this novel prey. They also licked the wet leg. If after the extensive grooming the test period had not yet passed, they walked along the walls of the test space while mewing.

Carrying. Carrying a mouse chiefly occurred after the prey had been caught and before it was bitten to death. It appeared as if the kitten wanted to find a place to kill it without being disturbed. A kitten often went to the same place on consecutive days, selecting corners which provided cover. Well-trained kittens did so less often. Sometimes a partly eaten mouse was carried, more often by younger than by older kittens. The circumstances of the arena also promoted this behaviour pattern. A big tomcat trained to catch fish, when confronted with mice in exp. P of Series I, caught the mice with the mouth but failed to kill them and spent the entire test time carrying them around. As mentioned on p. 57 fish are not killed before eating starts.

Touching. Touching the mouse with the forepaws is most common in p_2 where it occurs when more time passes before catching, and between catching and killing than is usual for skilled mouse-catchers. Much less often it is done with a dead mouse. In this behaviour the mouse is released from the mouth but kept under control with the paws while shoveling it on the floor. The higher percentage of touching in tests with older cats is again due to the higher incidence of inexperienced kittens in Series II. The two isolated kittens of exp. R did not even catch the mice with the mouth, but shoveled it with the forepaws more extremely stretched than we ever saw in experienced mouse-catchers (see Table

14). In contrast with the latter kittens they also drew their head somewhat backwards and directed the pinnae sidewards instead of forwards. When these kittens grew older the appearance of this behaviour pattern became more and more wild; they threw the mouse through the air, without using the mouth like the other kittens.

Table 9 shows that tests with touching were in kittens under 56 days of age more frequent in the presence than in absence of the mother (the same is indicated for tossing and rolling, but tests with these activities were much rarer). The mother, therefore, increases the probability that the kitten makes bodily contact with the prey during a test. Our data on the frequency of touching during a test, however, show that only in absence of the mother a kitten may continue touching for a long time. When the mother is present she usually takes over the mouse if the kitten repeatedly fails in catching and killing it, either to present it to another kitten or to immobilize or kill it herself.

Tossing. This pattern is in many ways similar to the foregoing one, except for the part played by the mouth. At the highest intensity of the activity the head is quickly thrown backwards while the mouse is released. Consequently the prey is thrown upwards, after which it is usually caught again with the legs and sometimes also with the mouth. In p_1 this behaviour was rare; in this period it could, because of the definition of catching, only be recorded when carried out without using the mouth. The largest number of tests containing this behaviour occurred in p_2; it was less frequent (with a dead mouse) in p_3 and rare during eating. Like touching it was more common in the kittens of exp. S. We shall argue below that this, in our opinion should be attributed to lack of experience in dealing with mice. However, this behaviour can also be promoted by various disturbances such as sounds outside or novel objects inside the experimental room. The incidence increased in kittens older than 280 days.

Head rubbing; Rolling. The data on these two patterns in Table 9 shows the same picture. For this reason and also because head rubbing against the floor is incipient to rolling we prefer to deal with them simultaneously.

Head rubbing plays a role in contact between cats, but kittens may also show it towards a mouse, touching it first with the nose, then with the cheeks and eventually sometimes with the body. In such cases mice may also be "handled" with the fore- and/or hind-paws. Because of the difference in size between mouse and kitten a considerable part of these movements did not actually bring the kitten in contact with the mouse but with the floor near to it. This is why rubbing often merged into rolling, the cat laying down during rubbing the floor on one side and then rolling over to rub the other side. Both behaviours occurred most commonly in p_2, were also relatively frequent in p_1, but much less common after the prey had been killed. They were far more frequent in the older kittens, particularly in those which had no previous training period with mice, such as the kittens of exp. S and the fish-catchers. When confronted with mice the latter kittens performed head rubbing and rolling with this prey until they had killed one and eaten it.

Head rubbing and rolling increased again in kittens older than 280 days to which a mouse was given (i.e. the *mr* litter). This increase coincided with a decrease of mice killing. The appetite of these kittens decreased and they often responded to one another with sexual behaviour. In this period (280-320) a number of kittens which had been given the opportunity of sexual intercourse became pregnant. Because of this change in the complex of factors underlying the behaviour of cats over 280 days old the comparative experiments on prey-catching were not carried through with animals after they had reached this age.

The most interesting result of this analysis is the increased occurrence of several interruptive behaviour patterns in kittens which have difficulties in catching, killing, or eating their prey. This is especially interesting because several of these interruptive activities, in particular touching and tossing, are often denoted as "play", which – depending on the author using this term – may have all sorts of implications with regard to the causation and function of such activities.

It seems to us that some important suggestions on the causation and function of these interruptive activities can be obtained from looking at some more details on the occurrence of this behaviour in the tests of exp. S in Series II. These kittens, after having been raised in the laboratory together with their nest mates but without experience with mice, were for the first time confronted with a living mouse at an age (varying within the litters) between 84 and 125 days. In each litter some kittens developed into successful catchers and some not (Table 11). In this table the data have been grouped for tests in which the kittens did and in which they did not seize the prey with the mouth. Because kittens only rarely failed in catching a mouse after they had once been successful each group mainly consists of tests with kittens that either did not or only scarcely fell into the other group, or changed groups after assuming the catching habit.

When comparing in Table 10 the occurrence of interruptive activities in each of both groups it becomes clear that "non-catchers" have scored higher on mewing, walking (without prey), grooming, and sitting, whereas "catchers" performed more growling, tossing, and rolling. Touching occurred in both groups, but in the "catchers" it often passed into tossing. The scores for rubbing cannot be considered different for both groups. The performance of rolling is influenced by the possibility for the kitten to bring the mouse in contact with its own body. Correspondingly it occurs more often in the tests of the "catching" group. However, when occurring in the other group it tended to reach higher frequencies per test. This is in accordance with our interpretation of rolling as a sexually motivated invitation to the mouse; as long as the mouse is moving freely such an invitation would be more likely to last shorter and to be repeated than when the mouse is kept under control.

To conclude this section on interruptive behaviour we express the opinion that the data obtained indicate that this behaviour is facilitated when a kitten is inhibited from catching a mouse; i.e. when it is in a conflict between the tendencies to attack and to escape (or avoid the prey). Some of the interruptive

Table 11. A survey of the success in mouse-catching of the 11 kittens of exp. S.

Litter and age when tested	Kitten nr.	Number of tests with:	
		Mouse not caught	Mouse caught
sg			
125	27	6	19
to	29	25	0
219	30	9	17
days	31	3	22
mi$_2$			
98	33	15	11
to	34	5	24
185	35	27	1
days	36	22	5
cy$_0$			
84	37	15	10
to	38	0	27
159	39	25	1
days			
Totals	11	152	137

activities occurring under these circumstances are not likely to influence success in later confrontations with prey, in particular grooming, walking, and mewing. Others, however, such as touching, tossing, carrying and rolling lead to contact with the prey and thus to the possibility of gaining experience. Growling and hissing, being agonistic activities, might facilitate prey-catching if their performance would have a stimulatory effect on the tendency to attack, a possibility which to us seems worth considering.

IV. General discussion

In this discussion an attempt will be made to combine the results of the more detailed quantitative study of the ontogeny of prey-catching (chapter III) with the mainly qualitative observations on the development of other – in particular social – behaviour patterns (chapter II) and with relevant data from the literature, to produce a synthetic picture of the morphogenesis of the behaviour of the domestic cat as an entirety. As already stated in the introductory chapter I, behaviour develops, like anatomical structures, not as separate groups of elements with a common function but as an architectural structure in the erection of which any element, irrespective of its ultimate role in the functioning of the individual, may act as an essential support to another element.

IV.1. THE DEVELOPMENT DURING THE FIRST MONTH

The behaviour patterns developing during the first month are mainly those which subserve almost any activity of the cat, such as the assumption of various body postures and locomotion.

In the development during the first month three phases have been distinguished. During the first one the lateral swaying movements of the body dominate; although the legs are able to move actively they give little or no support to the body.

This first phase recalls the observations of Preyer (1885), Carmichael (1926, 1927), and Coghill (1929) in particular on the development of the locomotory movements of amphibia-larvae. The earliest movement occurring in these larvae was a lateral alternated ⊂ – and ⊃ – bending of the body. It started anteriorly, in co-incidence with the formation of cross-connections between the sensory cells on one side and the motor cells on the other, and gradually spread to the posterior part of the body – in step with the development of new motor cells. When this movement was later on replaced by an undulating S-movement, Coghill found that new longitudinal connections between the sensory cells and the motor cells on one side had developed. Such changes in behaviour, due to growth of the nervous system, are usually categorized as maturation and this process is generally considered to take place independently of stimulation resulting from the performance of the movements, and in particular from their effect on the environment. Carmichael (1926, 1927) has shown this to be actually the case in tadpoles by immobilizing them during part of their developmental period by means of narcosis. Brought back to normal conditions their locomotory performance proved to be equal to that of larvae which had been continuously kept under such conditions.

For mammals the independence of a developmental process of adequate stimulation from the environment is much more difficult to prove. In the kittens the sigmoid swaying pattern and (provided the kitten is homogeneously supported) the leg movements co-ordinated with it, are available immediately from birth. This strongly indicates that these locomotory coordinations develop without an obligatory influence of their effect during adequate performance. Consequently for this development the term maturation seems applicable.

The sideways swaying of the head by which the nipple is located on the abdomen of the mother more strongly involves the anterior than the posterior part of the body, but is otherwise indistinguishable from the locomotory swaying movement. It is also available from birth on and it seems neither released, maintained, or shaped by specific characteristics of a substrate. It possess the characteristics of an automatism (p. 16). Only its discontinuance depends on specific stimuli from the nipple.

In the second phase the legs become gradually strong enough to support the body. Statokinetic postures can be maintained and in addition to the lateral movements of the first phase, dorsoventral movements become possible. As a result of these processes, which could possibly also be labelled as maturational, at the end of the second phase the kitten is able to stand and to walk.

The third phase is characterized by the emancipation of the legs. This means that the kitten no longer needs all its legs to maintain equilibrium and thus can use one or two legs for other purposes, such as scratching or grooming its body, but also the manipulation of objects in the environment. Thus, one may say that at the end of the 4th week the kittens are in command of their motoric equipment. As in the 3rd week the sensory organs have also become functional the kitten is sufficiently implemented at the end of the first month to enter into contact with environmental situations beyond the mother and the immediate surroundings of the nest. At that time the first rudiments of prey-catching can be seen, and, although interactions with littermates already occurred while striving for a nipple, they take place from then on independently of the mother and often appear to be undertaken intentionally. Although we shall now discuss the development of prey-catching and of social behaviour separately, one should keep in mind that this separation is artificial. As stated in the beginning of this chapter, the patterns of both functional groups develop simultaneously and experience obtained in one context is likely to influence the other.

IV.2. THE DEVELOPMENT OF PREY-CATCHING

When the kittens are four weeks old the movements of the jaws used in killing a prey and enough movements of the forelegs to manipulate prey are available to the kitten. At this stage these movements are applied with all the physical strength the kitten possesses; its nails are fully extended and the jaws are used with maximum force. One kitten even caught and killed a mouse at day 27 (p. 36). Table 1 shows that several other kittens killed mice when they were about one month old and most of them before the age of 6 weeks. The youngest age at which Yerkes and Bloomfield (1910) saw a kitten killing a mouse was 28 days; they observed it several times in kittens between 28 and 56 days old, and so did Kuo (1930). These findings are not in accordance with Leyhausen's (1957[a,b], 1965) statement that the killing bite (Tötungsbiss) is always the last prey catching activity which appears. Leyhausen is of the opinion that this late appearance is due to late maturation (Reifung). As the killing bite can appropriately be performed without earlier experience with mice, we agree with Leyhausen that it can be said to develop through maturation. However, we disagree with the statement that it would mature later than other elements of predatorial attack. Furthermore Leyhausen writes that the first occurrence of this killing bite is usually facilitated by some additional environmental stimulus, not directly emanating from the potential prey and increasing the aggressiveness of the kitten (e.g. the approach of a competitor). The same observation was earlier made and reported by Berry (1908); we too have noticed it. The fact that the approach of a competitor can facilitate the attack on a prey, indicates that the social and the predatorial attack have motivational factors in common. Moreover, as will be set forth below, the basic form of the attack pattern is similar in both cases. It seems likely to us that here the same behavioural mechanism is serving different functions.

The facilitatory effect of particular external stimuli on the release of a predatorial attack provides a clue for explaining the discrepancies in the statements about the age at which a first killing can take place. Although the motor patterns for grasping and killing are available to a 4 week old kitten, it depends on the relative strength of its motivation for attack whether and when the first execution occurs. Stimulation of the attack system – by any internal or external factor – facilitates the occurrence of grasping and killing. In contrast activation of escape may inhibit and postpone these activities. This happens when a potential prey intimidates the kitten – because of its relative size and/or the fierceness of its defensive behaviour. In addition the threshold for escape may also be influenced by earlier experience.

The mouse we saw killed by the 27 day old kitten was a small specimen; we would not expect the physical force of the jaws of a one month old kitten to be sufficient for killing a big mouse. However, when a kitten has reached the age and size corresponding with the physical strength necessary to kill a big mouse it is – as will be explained below – also more likely to be on the average more inhibited in its approach behaviour (thus have a lower threshold for escape responses) than a naive 4 weeks-old one. This effect will tend to further postpone the date of the first killing. The additional aggressive stimulation mentioned above will counteract the inhibiting tendency to escape.

As soon as a killing has been successful and the kitten has eaten from a prey the probability that the next prey offered will also be caught and killed increases dramatically (as illustrated in Fig. 9). One might say that the kitten at that moment recognizes the use of the already available pattern and has consequently become able to put it into practice.

However, in natural situations the act of catching and killing cannot be performed before the prey has been searched for, detected, recognized, approached, and brought within reach. Moreover, the bite has to be applied in such a way that it is effective. The development of the complex ("appetitive") behaviour through which all this is achieved takes – as shown in chapter III – more time than the maturation of the killing bite. The various building elements of the appetitive behaviour, such as the different possibilities of locomotion, watching postures, stalking, and pouncing are not yet linked into effective sequences when the kittens are one month old. The development of the full appetitive behaviour not only takes more time it also often shows in its form individual variations between different kittens. The possibility demonstrated in this study of modifying predatorial behaviour in individuals through particular manipulations of the external situation during the developmental period shows that processes other than maturational ones (in the sense as defined above) must be involved.

To catch and kill a prey efficiently the kitten has first to identify an object as a prey and to assess its catchability. Further, it has to decide about its tactics concerning the situation in which the prey occurs; this involves the way of approach and the type of manipulations to be used. Finally, it has to choose the right moment to grasp the prey, to orient itself properly with respect to it and to apply the correct force. In the following we shall first look at the indications we obtained for influences of experience on the development of these aspects of skilled behaviour.

Identification of the prey. As to the question of how far knowlege of the suitability of different species as prey is obtained from experience with such animals, in particular when caught by the mother and brought to the nest, no conclusive answer was obtained. The parallel experiments in which either mice (exps. M and N) or fish (exp. P) were offered to the kittens in the early period were aimed at answering this question, but in this aspect they failed since the tactics of dealing with a prey under water dominated the results of these experiments.

All authors agree that a rapidly moving object is a strong releaser for predatorial behaviour even in unexperienced kittens. Furthermore kittens are very alert to gnawing or stridulating sounds. Fish odour is always very attractive to cats. On several occasions we have seen in 3-5 week old kittens, unacquainted with fish, strong positive locomotory responses towards a non-visible source of fish smell. We only succeeded to arouse similar responses with mouse odour in kittens experienced with mice. Berry (1908) states that kittens inexperienced with mice only became interested in dead mice after the mother had opened them up and exposed the flesh. In this way all normally raised kittens will have opportunities to become acquainted with the specific smell of the prey animals brought home by the mother. The importance of this experience in later life was shown by Kuo (1930) with regard to eating rodents. He found that 80% of the kittens he raised on a vegetarian diet later refused to eat rats.

Exp. T, in which during the early phase the kittens had the possibility of roaming about in the garden, gave them the opportunity to learn which objects are not suitable as prey. Animals like frogs, toads or slugs are mainly touched with the paw and barely with the mouth, obviously because of the repellent effect of their skin. In contrast, shrews are caught with the mouth, carried around, brought to the nest, killed and sometimes at least partly eaten. However, the latter only happened in young kittens, and in all cases in which eating was established with certainty, the kitten vomited after fifteen to twenty minutes. The kittens, and also older cats, may manipulate shrews for long periods without killing them. The shrews may stay for a long time in the neighbourhood of the cats, hiding in crevices then turning up again. They obviously stimulate hunting behaviour in the cats because of their quick movements and high pitched squeaking. The harrassing may eventually cause the death of the shrews, but we have never seen older cats eating them. Apparently specific chemical features of the shrews only become repellent at close oral contact. Mice and voles are never hunted for such long periods; they are either rather quickly killed and eaten or they manage to escape.

Kuo (1930, 1938) had demonstrated that experience with a potential prey during the nest-period can also lead to learning not to catch it. When he raised kittens without their mother from a very early age on together with albino rats, a majority of such kittens did not catch or kill albino rats, not only during the period they were living together with rats but also afterwards. In contrast they did catch individuals of pigmented strains of rats and mice. This selective abstention must have been the result of experience during the nest-period. Two different kinds of experience have to be taken into account: (1) as the rats were adult by the time the kittens were able to perform predatory behaviour they are likely to have successfully defended themselves and scared off the kittens; (2) the kittens may to some extent have socialized with the white rats. The latter could be proved for single kittens kept with an albino rat (Kuo, 1938); the probability that a kitten familiar with white rats would in the long run kill such rats was found lower than in a kitten without this experience.

Kittens kept with white rats did eat the naked young of these rats if the opportunity arose. Such kittens later on killed shaven white rats, neglecting unshaven ones, but so did one of 5 kittens which had never killed newly born rats.

Catchability of the prey. Prey animals which are considered to be oversized with regard to the capacities of the kitten are likely to be ignored. Further, the size of the prey may influence the catching and killing tactics chosen. This is because experience with a prey effectively defending itself by attacking and biting the kitten, influences its later behaviour towards the prey. Especially in cases where the aggressiveness of the mouse had increased during the encounter and become more effective, kittens for several days were unwilling to catch the mice presented to them. They were also touching and beating the aggressive prey more often with the paws. These observations are comparable to the unlearning obtained by Kuo (1930) when he stimulated a kitten while catching a rat with an electric shock. Relevant in this context is also the effect the anomalous mouse-catching behaviour of mother *cy* had on her kittens. This cat had as a kitten in our training procedure assumed the habit of thrusting herself on a mouse in an extremely fierce manner. She did so also in the presence of her own young who as a result got scared, and when a mouse was put into the cage crouched and consequently tended to keep away from mice.

With regard to the effect of prey size, we undertook an experiment in which white rats of three different sizes were presented to the four kittens of one litter *(mr_l)*. A survey of the results in presented in Table 11. The following protocol may serve to illustrate the proceedings during these tests. It concerns the presentation of a white rat of 250 gram to *mr_l-1* (6 month old).

Oct. 19, 1967. 10.50. Kitten *mr_l-1* is introduced into the arena in which a white rat of 250 g moves freely around. Kitten approaches the rat and sniffs at it. Consecutively it performs rolling 4 times in the middle of the arena. The rat is walking around and the kitten begins to follow him slowly, without touching him. When a moment later the rat approaches the kitten the latter withdraws and rolls again at 2 m distance from the rat.

10.52. Kitten touches rat for the first time, snaps in the direction of the neck of the rat. Rat squeaks, kitten beats it with paw on its back. Then the rat approaches the kitten, which slowly withdraws while fiercely slamming its paw at the rat. The rat continues walking in the direction of the cat whereupon the latter retreats and starts grooming its hindlegs. The rat advances further towards the cat, which slams at it 2 times. The last slamming movement merges into washing the paw. The cat moves away from the rat to the centre of the arena and continues grooming, this time its flank. A moment later the kitten again approaches the rat, does not touch it and walks away. Far away from the rat the kitten repeatedly vocalizes: mrauw.

10.55. Kitten washes hindlegs and anal region. Rat runs along the wall of the arena; the kitten follows it with the eyes. When the kitten gets up and begins to follow the rat again, the latter turns and approaches the cat, who is silent as long as he is concentrating his attention on the rat.

10.57. Rat runs along the walls, the kitten following him at 20-30 cm distance. A number of times the kitten moves its head near to the rat, however, without making snapping movements. Then it goes again to the centre of the arena and starts grooming and rolling alternately.

11.00. The door of the arena is opened and the experimenter tries to induce the kitten to come out. The kitten "obeys" reluctantly, after having once more approached the rat and groomed herself.

Table 12 shows that the bigger the rat the less complete is the attack. Further a stronger kitten shows a more complete attack against a big rat than a weaker one. If the attack is inhibited rolling tends to occur. Kitten *mr_l-2* even performed this behaviour after it had killed the 400 g rat, following an immediate pounce and bite in the neck. The kitten then rolled on its back near to the dead rat before it started eating. Grooming always took place when the kitten appeared to have given up attacking the rat (see also p. 65).

Table 12. The behaviour of the kittens of litter mr_l towards prey of different sizes. The kittens had been trained to catch mice from the 4th week on. On day 182 white rats were presented to each kitten when alone in the arena. The experiment started with the rats of 400 g and were after one hour continued with the rats of 250 g. Again one hour later a rat of 140 g was given of mr_l-*1*. The estimated strength of the kittens increases from left to right.

Weight of the rat	140 g		250 g		400 g			
Kitten nr.	mr_l-*1*	mr_l-*4*	mr_l-*1*	mr_l-*3*	mr_l-*4*	mr_l-*1*	mr_l-*3*	mr_l-*2*
Killing	+	–	–	+	–	–	–	+
Biting rat	+	–	–	+	–	–	+	+
Snapping at rat	+	+	+	+	–	+	–	–
Touching rat with paw	+	+	+	–	+	–	–	–
Rolling	–	+	+	–	+	+	–	+ after rolling
Grooming	–	–	+	–	+	+	+	–
Ignoring after x min	0	8	3	0	5	5	6	0

Tactics. The differences we found between the ways of approach of the kittens who had experience in the garden during the early-period and those who only received their experience in the laboratory training tests, reveals an influence of experience and learning on tactics. The former group must often have encountered potential prey which escaped when not aproached cautiously. In contrast during training in the laboratory there was not much possibility for the mice to escape. The "garden-kittens" were always stalking during their approach; in contrast the mouse-trained laboratory kittens rushed towards the prey. Kruuk and Turner (1972) found that in cheetahs of the Serengeti plains stalking, although present in the cubs, rarely occurs in adults apparently as a result of experience.

Compared with the trained mouse-catchers the fish-catchers showed a difference in tactics when trying to get a mouse. Just like the mouse-catchers the fish-catchers had no difficulties in reaching a mouse at the bottom of a dry aquarium, but only they initially often "angled" the mouse from the aquarium with a leg and extended nails; later on they did it with the mouth only, just like the mouse-catchers. Unlike the mouse-catchers the fish-catchers were not accustomed to biting their prey to death before they started eating; although the fish might sprawl on the dry substrate, they did not run away like mice. This was probably the reason why with the fish-catchers more mice managed to escape than with the mouse-catchers; the average time between catching and killing was also longer in the former group. Although some fish-catchers rather quickly adapted the habit of killing mice, others – among them the biggest tomcat in this group – needed more than twelve tests for it (Table 7b). When later on one of the fish-catchers got pellets of foodcake in milk, it angled the pellets from the bowl, after which it took them from the paw with the mouth, just as it used to do with fish.

Although fish is an attractive prey for all cats, independent of experience with that kind of food, only one of eight kittens trained in the early period to catch mice and none of all four cats who obtained experience in the garden were able to catch a fish in an aquarium filled with water. Here not the kind of prey had to be learned, but the kitten had to learn how to overcome the difficulty of taking it from the water.

Twice we have come across a housecat who had taught herself an individual habit for obtaining fish. One of them angled baitfish from a cauf which hung in the water, by placing herself on top of it and angling with one paw through a hole in the cover. The other took position near fishermen angling from the bank of a canal. Every time a line was swung out of the water, the cat jumped upon the fish caught. As the cat was usually allowed to keep undersized specimens, its behaviour was frequently rewarded.

The mode and orientation of grasping the prey. Cats experienced with prey of different sizes tend to use only the mouth for grasping small prey. With somewhat bigger prey they may use a leg in addition. In encounters with rabbits our cats used both forelegs.

With mammalian prey the killing bite is directed to the neck. Leyhausen (1956) has experimentally demonstrated that adult cats grasp a prey just behind the head and he has given arguments that this response is visually directed. For opening up and eating, which also starts at the head, the cat would use tactile information on the direction of the hairs as a cue. In a later paper Leyhausen (1965) revises his earlier opinion that the direction of this orientation would be "fully innate". Without giving further experimental evidence he is restricting this statement to the direction of the movement towards the area behind the head, and believes that from this basis the correct grasp of the neck develops through learning. Eisenberg and Leyhausen (1972) in a comparative study on the extent to which various predatorial mammals and marsupials use their paws and mouth, found a correspondence between the involvement of paws and claws and the amount of opposition the predator is likely to receive from its usual prey species. These authors are of the opinion that the specialized claws of the Felidae, as well as the occurrence of canines in the denture, are adaptations for catching large and difficult prey. They also believe that the more difficult the prey species the more elements the predator has to incorporate during the development of its catching patterns. Eibl-Eibesfeldt (1956, 1963) on the basis of film analysis and experiments came to the conclusion that in the polecat learning is involved in the ontogeny of the neck-bite. He suggests that in this respect not only experience with prey animals but also playing with nest mates might be of importance.

In our experiments improvements in orientation with increased experience were in particular obvious in the kittens of Series II, who had no encounters with prey in the early period (exps. R and S). When confronted with mice these kittens in the beginning often seized them at the legs or the tail and only gradually directed their attacks towards the neck. Not many trials, however, are needed to improve the orientation. In particular kitten *mr2-13*, one of the kittens reared in isolation for exp. R. who for the first time began to catch mice after she had become a mother *(mo)*, had difficulties in orientating herself towards a prey. It is obvious to relate these difficulties at least partially to the fact that these kittens could not interact with littermates, while other moving objects in their environment – like those met by the kittens reared in the garden – were in exp. R also absent in the laboratory cages.

The fish we offered were never killed with a neck-bite, although they were always grasped dorsally with the mouth. This, however, is the obvious follow up of the angling movements; the swimming fish is pushed with the paw against the wall of the aquarium and moved upwards, its back is the part the cat can reach without wetting is nose. Depending on which part is nearest to the mouth of the cat it grasps the back rostrally or caudally; we have no indications that the bite is preferentially aimed at the area just behind the head of the fish. Next the fish is laid on the ground and kept down with a paw. Then eating starts at the gills and only as a result of this the fish is killed. The cats probably find the gill area by touch and select it because of its relative weakness as compared with the rest of the other scaled parts of the body.

The force applied in the activities. An interesting phenomenon, which has bearings on all movements carried out to catch a prey and to which Lorenz and Tinbergen (1938) have explicitly drawn attention in their analysis of egg-rolling in the grey-lag goose, is the adaptation of the amount of force invested in activities of the fixed pattern type to the objects normally releasing them. The authors mention that this will often be brought about by a balanced interference of antagonistic muscles, but they also present evidence that in the sagittal movement of egg-retrieving the extra energy expenditure involved in such a balancing mechanism has apparently been economized by predetermining the force with which the fixed pattern is carried out. Objects considerably lighter or heavier than normal eggs disturb the form of the movement. The authors do not discuss the causation of this effort constancy; one gets the impression that they tended to incorporate it into the "innateness" of fixed action patterns. The adapted assessment of force to an activity is typical for various components of prey catching behaviour; it evidently improves, however, while the kittens are gaining experience.

IV.3. THE PROCESS OF GAINING EXPERIENCE

The results of the experiments have shown great differences in the time necessary for becoming a skilled preycatcher and in the nature and the sequence of stages passed before the ultimate pattern was reached. This variation was found to depend on the situation in which the kitten grew up; some determinants involved could be experimentally manipulated. However, also under constant external conditions the variation between kittens was still considerable. This is in particular striking in exp. S in which only part of the kittens who obtained their first mouse at an age of at least 12 weeks (thus in the later period see p. 45) became mouse-catchers afterwards. This stands out strikingly against the case of kitten *n_l-11* who caught and killed its first mouse in the early period, when it was 27 days old.

Table 13. Survey of main occurrences in all tests with the kittens of the S-series.
a: catching took place in tests nr. 11, 12, 14 and 15.
b: no grooming in the 4 tests in which a mouse was caught.
c: seizes mouse once in test nr. 1, but releases it immediately.

Litter nr.	*cy-0*			*mi-2*				*sg*			
Kitten nr.	37	38	39	33	34	35	36	29	30	31	27
Age when first mouse was presented	84	84	84	98	98	98	98	126	126	126	140
Number of mice presented before first mouse was killed	–	0	–	12	5	–	–	–	9	6	7
Age when first mouse was killed		84		147	110				148	140	158
Number of mice seized or bitten before first mouse was killed		0		1	1				0	2	1
Number of mice seized or bitten without mouse being killed	4[a]		1[c]			0	0	0			
Kitten attacked by mouse	–	x	–	–	–	–	–	–	x	–	–
Mouse quickly escapes	–	–	–	–	x	–	–	–	–	–	–
Number of presentations	18	19	19	16	16	17	16	16	16	16	16
Number killed	0	19	0	3	11	0	0	0	7	10	9
Number of tests with grooming	10[b]	0	14	4	2	9	5	7	2	0	1

It is interesting to compare the case-histories of the 11 S-kittens during the experimental period. Table 13 has been composed for this purpose. It shows that the number of mice presented before the first one was killed varied between 0 and 12. Moreover 5 kittens did not kill a mouse in 16-19 presentations. Of the kittens who killed a mouse 4 already had caught one the day before. Of the non-killers kitten nr. *37* caught a mouse in 4 tests at the end of the series; it is likely that it would have become a mouse-killer had the series been continued. Thus, in a number of kittens (*33, 34, 31, 27* and perhaps *37*) within 2 or 3 days an internal change seemed to have taken place which made them attack the potential prey. The other two mouse-killers (*38* and *30*) were probably lifted over the threshold through external stimulation; both were attacked by the mouse and retaliated (see p. 69). After the first mouse had been killed the chance that following ones were killed was practically 100% (only kitten *33* once did not kill a mouse within the test period but only carried it around). When interruptive activities occur in the test the chance that the kitten will catch the mouse is low. This is shown by the occurrence of grooming in 45 tests in which the mouse was not caught or killed and only in 5 tests of kittens who became mouse-killers; i.e. in 5 of the 39 tests before the first killing. The phenomenon of kittens developing divergently either into mouse-killers or non-mouse-killers reminds one of Hogan's (1965, 1966) description of the chicks which eat and which refuse a mealworm. We are – in accordance with this author – inclined to explain the phenomenon as the result of the establishment through a conditioning process of different attack-escape equilibria, specific for a particular external situation (i.e. the prey species involved). Chicks that do not eat the mealworm often run around with it, uttering alarm calls. Kittens also carry around mice for longer periods when signs of being frightened by the prey or other stimuli in the surroundings are present.

In spite of this considerable variability in the course of the development the final forms of the different predatorial acts differ – at the level of exactitude we have been making our records*) – surprisingly little between individual domestic cats. This is true for mouse catching, in which the prey is approached from behind, seized with the mouth behind the ears and quickly killed, all with not more than little assistance from the legs. It is just as true for fish catching, where in the final form the fish was taken out of the water by moving it with one frontleg against the wall of the aquarium; caught with the mouth as soon as it was above the water level and then carried to a quiet site for eating. The angling pattern was also used for fetching a mouse from a crevice. Also the seizing of large prey with both legs had the same form in different individuals. This form is similar to that of the pattern by which lions catch big mammals (Schaller, 1972; Leyhausen, 1957[b], 1973) but lions, when dealing with small prey, also use the pattern involving the mouth only (pers.obs.).

The end results of the processes through which these patterns emerge can therefore be called species-specific or, with Leyhausen's comparative studies on various Felid species in mind, even taxon-specific. It seems as if during the developmental period, at least in the normally living kitten, opportunities are provided for the performance of motor patterns already available to it. When as the results of performing a particular pattern the kitten perceives a specific reinforcement this activity appears to have found its place in the sequence of elements of predatorial behaviour and starts to be applied adaptively in this functional context. The question to which extent the sensitivity for this reinforcement is based on genetic and on acquired information must at this moment remain unanswered. Ideas similar to those expressed here can already be found in Kortlandt (1940) in relation to the nesting behaviour of the cormorant.

The road to this first performance is roughly programmed ("guided" development). To begin there is the period of "play" with the mother, littermates and small moving animals, as normally available sparring partners. Further, although our data did not reveal clearly limited sensitive periods, we found some indications for a certain timing of the possibilities for learning the different ways of catching: it was impossible to teach kittens fish catching at ages below 50 days, when mouse catching was already possible. The striking difference in variance of the time needed to become a mouse-catcher between the exps. M and N on the one hand and exp. S on the other, must be due to the effect of the training period on the M and N kittens. Our training efforts are comparable to the influence of the mother when bringing prey to the nest. It is not unlikely that the kittens would be sensitive to experiences with prey in

* The possibility should not be excluded that very exact and detailed descriptions as can be obtained by means of choreographic methods (e.g. Golani, 1976) would still reveal important individual differences of the performance of what here are considered as identical acts.

particular at the time of weaning when they would need novel kinds of food. The observations of Yerkes and Bloomfield (1910) and of Kuo (1930, 1938) discussed on pages 7 and 69 of this paper, respectively, point to such a sensitivity in the early period.

Although as yet insufficient data are available we consider it likely that the basic motor pattern of each of the different acts applied during predatorial behaviour, develops mainly through maturation, i.e. largely independently of external stimulation and in particular of feedback information about their effects. Weiss (1952) reports that they were normally developed in 2 kittens he raised in isolation from conspecifics. Thomas and Schaller (1954) found that optically deprived kittens (wearing opaque glasses from the first week on) were nevertheless able to perform the motor patterns of prey catching when they were, at the age of 11 weeks, confronted with a simple moving prey dummy in a simple experimental standard situation. However, the fact that these activities possess the character of fixed action patterns does not exclude a role of learning in their application, as the latter authors suggest.

In contrast to the stereotypy of the form of the separate motor acts, a considerable amount of variation between individuals occurs in the frequencies in which the different activities are applied and in the sequences in which they are linked together, i.e. in details of the "appetitive behaviour", such as the occurrence of stalking in exp. T. Individual differences were also observed in preferences for certain species of prey and in the degree of skill exhibited. Years later cats still showed idiosyncrasies in their predatory behaviour which they had developed as kittens during the experiments. Correspondingly, West (1974) found individual idiosyncrasies in play behaviour of kittens.

Under normal conditions the mother will influence the development of predatory behaviour of the kittens by bringing them into contact with one or more kinds of prey and by giving them an opportunity for practising with prey specimen she has made groggy. However, the mother is not absolutely necessary for the development of catching and killing of prey. Between the fourth and the eighth weeks her presence may even slow down progress in the kitten when she interferes by killing the prey herself after one or two failures of the young. Presence of nestmates may cause interference until the prey is caught. From then on, however, interference by nestmates facilitates the occurrence of a strong killing bite, perhaps as a result of an increased stimulation of the tendency to attack caused by the rivalry in that situation (see p. 44). Kittens do not interfere as long as a littermate is eating a prey. However, when a kitten leaves part of its prey, a littermate may take it over. This is in interesting contrast with the finding by Rasa (1973) in the dwarf mongoose (Helogale undulata rufula). In that species fierce competition over food occurs between littermates. We are inclined to bring this difference in connection with the difference in claw structure between the Felidae and the Herpestinae; we consider the "protractile" claws of the former as a more dangerous weapon (hooks shot into the tissue of the opponent) than the claws of the latter which have no protrusion mechanism (Ewer, 1973).

We agree with those authors (see section 5 of this chapter) who consider "play" as a very important vehicle for gaining experience relevant in later life. Egan (1972, 1976) has made a study of "object play" in kittens. She has described the different patterns involved, classifying them into categories such as sniffing, paw-play, mouth-play, crouch and pounce play, and rolled-play. In particular she has paid attention to the releasing effect of qualitatively different objects for the various play activities. For this purpose she presented to the kittens large (14 cm long) and small (9 cm long) sausage shaped objects of real fur and of nylon "fur", a newspaper ball, a woollen pom-pom, a bald sausage shaped object of cured-skin (one with and one without strong smell) and a dead mouse. The ball and the pom-pom were given free from the floor and hanging on a string. In general the stimulus strength for releasing play (measured as relative frequency of play patterns) decreased in the order: dead mouse → small real fur, small nylon fur → smelling bald skin and swinging pom-pom or paper ball → large fur, non-smelling bald skin and pom-pom or paper ball on floor. The sound of a moving object also played a role.

The object was usually first sniffed and/or patted. The kind of responses following largely depended on the stimulation further received from the object. Fur tended to release biting and biting facilitated shaking, throwing and carrying. Crouching and pouncing were in particular triggered by movement of the object. The experiments cannot answer the question in how far the effectiveness of the different objects in eliciting play is influenced by experience. All kittens had experience with the real fur of mother or littermates; 15 minutes of confrontation with a hitherto unknown object proved to be sufficient to bring the level of playing with it to the maximum obtainable for that object, irrespective of whether this first confrontation took place when the kitten was 20 or 8 weeks old (or even younger).

After an increase of the frequency of play with the different objects from the 4th to the 8th week the time spent playing remained almost constant until the 24th week for the small real and nylon fur object, but it decreased for the less attractive objects. Almost no changes with age took place in the relative frequencies of the various play patterns as long as the external conditions remained constant. However, the type of play was influenced by the presence and interference of littermates and as the social relations change with time in groups of kittens, the relative frequencies of different types of play are affected. Because they are potential rivals the presence of siblings can restrain the occurrence of certain play patterns in a kitten; on the other hand in general their presence stimulates playing.

In our opinion a kitten when moving towards a prey is motivated by more than merely a tendency to move in a certain direction. This is why we have intentionally spoken of a tendency to attack instead of a tendency to approach. The locomotion is accompanied by non-locomotory patterns such as snapping at the prey, beating it with a paw and growling. We consider these patterns to be under common control of a specific behaviour system: the attack system, but we are aware that an extensive causal analysis of the entire behaviour of the domestic cat would actually be necessary to prove this point.

The reluctance in attacking big prey, and perhaps also a part of the orientation towards the hind parts of the prey, are influenced by a tendency to flee. Overt indications that this tendency is activated during such occasions, such as pinnae laid down against the head, backwardly directed whiskers, and half closed eyelids, can be observed. As mentioned on p. 20, such elements of fear were first seen in the 3rd week. The complex of causal factors underlying these patterns as well as the hesitation to attack we have for the sake of brevity called the "tendency to flee or to escape". We realise that much more causal and analytical work would be necessary for a satisfactory justification of this term, particularly since in some animals indications for the presence of different, even competing escape systems have been found (Hogan, 1966; Baldaccini, 1973).

A tendency to flee is, in our opinion, of great importance for the development of the behaviour of the kittens. We believe that it is the interference of this tendency with the tendency to attack the prey that causes the delay in which so called "play" can occur. As shown in chapter III during this delay interruptive activities are carried out, among which touching the prey with the paws and tossing it with the mouth are frequent. In particular these activities, combined with walking and running, are the main components of what generally is called play in cats. Our conviction that the tendency to flee is important in creating the opportunity for these behaviour patterns is mainly based on the fact that their occurrence can be induced by slightly scaring a kitten. Experienced mouse-catchers that had acquired a habit of immediately attacking the mouse when presented in the familiar situation showed "play" when at the moment of presentation novel objects were around.

It is possible that factors other than fear which are able to inhibit attacking and/or eating a prey have effects similar to those mentioned above. We once saw this behaviour in two adolescent cheetas who, judging from the size of their bellies, must have been satiated, when they came across the fawn of a Thomson gazelle. We could not, however, induce such play in 10 month old kittens by offering them up to 7 mice successively, which with a total weight of 200 g amounted to the maximum portion the kittens at that age would take. In this experiment a mouse of 30 g was presented to kittens who were already well fed, immediately after a preceding one had been devoured. In 10 sequences of presentation with different kittens from 2 up to 7 mice were eaten successively. No differences were found in the speed at which the mice were caught and killed in successive tests of one sequence; eating only went slower in case of the 5th, 6th and 7th mouse. Then the kittens from time to time licked their paws and stretched themselves, as if they had difficulties in swallowing the food.

Indications of a tendency to escape were more frequent in the tests with the two isolated kittens of the R-series than in any other series of tests. Table 14 gives a survey of the incidence of the main activities recorded during the tests at different ages. Raising of the hair of the body and the tail, a very high incidence of touching, and wild jumping or charging not directed at the mouse or any other obvious target are only common in these particular tests. Kitten *mr*$_2$*-13*,

Table 14. Incidence of various behaviour components in R-kittens during tests with mice at different ages (+ = less then 10 times/10 min., + + = 50%, + + + = 100% of test period). 1) Seizes mouse 3 times in back but releases it immediately. 2) Rolls with mouse.

Series R	Kitten *br-21*						Kitten *mr*$_2$*-13*						
Age in days	Fur raised	Continued watching from distance	Pursuing	Nosing	Touching with paw	Vacuum charging	Age in days	Fur raised	Continued watching from distance	Pursuing	Nosing	Touching with paw	Vacuum snapping
102	+	+				+							
106	+	+				+							
109	+	+											
113	+	+				+							
120		+				+	120	+ +		+	+	+	
127		+				+	127	+ +		+	+	+	
134		+		+			131	+		+	+	+ +	
141	+			+	+								
148	+	+			+	+							
155	+			+	+	+	156	+		+ +	+	+ +	
162				+		+							
169			+ +		+	1)	169	+		+ +	+	+ +	
							176	+		+ +	+	+ +	
							198			+ + +	+	+ + +	+
							204			+ + +	+	+ + +	+
							211			+ + +	+	+ + +	+ 2)

which seemed to get nearer to catching a mouse than *br-21*, in the last tests repeatedly snapped its jaws beside instead of into the mouse as if a high attack tendency was at the last moment redirected; when this occurred the hair was no longer raised.

IV.4. THE DEVELOPMENT OF SOCIAL BEHAVIOUR

When at the end of the first month the kittens are capable of moving around they no longer struggle with littermates by only pushing them away, but begin to launch direct attacks upon each other. They first do this with full force, stretched nails and fierce bites, but after some encounters in which the opponent has answered the attack with similar behaviour, their approach of nestmates becomes more inhibited. The nails are retracted when beating a littermate and biting tends to be superficial (both in contrast of what in that period can be observed in the same motor pattern directed towards prey animals). When counter-attacked the kittens show overt signs of a tendency to flee.

The versatile mutual "chasing-wrestling-fleeing" pattern, which is so typical for littermates between the 4th and 8th week and for which in particular the term "play" is commonly used, can thus be seen as the consequence of oscillations of the levels of the tendencies to attack and to flee in the opponents.

It has been suggested above that through this "play" behaviour learning processes, important for survival in later life, might become established. An influence of the presence of littermates on prey-catching has indeed been found (p. 44); in particular influences on social behaviour are to be expected. A way to check this expectation is to deprive kittens during the periods concerned from experience that can be considered relevant in this context. It is imaginable that such influences would affect in the individual: 1) the thresholds for the control systems for attack and escape; 2) the development of communicative more or less stereotyped activities, wholly or partly controlled by these systems in correspondance with the "conflict hypothesis"; 3) the release and the orientation of such activities in general, and in particular their triggering by different specific stimulus situations.

The thresholds of attack and escape. Deprivation experiments of this type have so far been done in a few species: the red jungle fowl *(Gallus spadiceus)* by Kruijt (1964, 1971) and the rhesus monkey by Harlow and his co-workers (e.g. Harlow, 1969; Harlow and Harlow, 1962, 1965; Harlow, Harlow and Hansen, 1963). To our knowledge, for cats only some data are available from two kittens isolated from conspecifics by Weiss (1952) at the age of 6 and of 2 days, and from the two kittens of our exp. R; isolated from conspecifics after having been alone with the mother for the first seven weeks.

In the fowl and the monkeys the threshold for attack and escape were indeed found to be affected. When confronted with a social or sexual partner the isolated animals had great difficulties in controlling these tendencies in an adaptive way. They did not seem able to maintain a certain balance between those tendencies for some length of time; fits of excessive aggression and of dominating escape irregularly occurred in alternation.

When confronting one of his kittens with a conspecific after one year Weiss only obtained strong escape responses. The behaviour of our R-kittens was in several aspects aberrant from that of normally reared cats. In the R-kittens spontaneous biting at parts of the cage was more often seen than in kittens kept together with their littermates. When the caretakers placed dishes with food and water in the cages of the R-kittens their hands were more often bitten and scratched then when dealing with socially reared kittens. When walking or running – as in the mouse-catching tests – we much more often heard the nails tickling on the floor than in non-isolated cats who usually kept their nails retracted. Although the R-kittens sometimes directed attacks towards their mother, as long as she was still with them, they did not get much opportunity for biting since the mother used to catch them under her forepaw and while licking them induced them to suckle.

The two kittens of exp. R were not able to perform the meeting ceremony (p. 23). When approached by a kitten of another litter they retreated. As a result of this, after the other cat had approached and passed them a number of times in succession, they retreated backwards until they became cornered against the wall of the arena in which the confrontation took place, assuming the arch with neckflex. A further approach of the strange cat released biting and fierce beating with a forepaw in the isolated cat, whereby the other cat got scratched. The latter at most replied with a casual slamming and continued its walk. However, when the two met again, the isolated cat usually did not make a stand but fled towards another part of the arena. In an encounter of experienced cats of the same age such an overt appearance of attack and fleeing is much less likely.

Kruijt (1964) found that red junglefowl cockerels developed into harmoniously behaving cocks when in the first half year they had the opportunity to fight with other young males in their group. It seems likely to us that the wrestling of kittens after the fourth week serves the same effect. The development of an adaptive control of the balance between the tendencies to attack and to flee was in the jungle fowl and in the rhesus monkeys also important for the normal performances of sexual behaviour. This seems to hold for the cats too. When the R-kittens came into heat, they vocalized and rolled as normal cats in oestrus do. However, when they were approached by a sire they withdrew, hissed and scratched. In contrast normally a cat in heat meets the sire frontally and although this usually leads to some agonistic behaviour in which the sire may withdraw or flee, when he is persistent without proceeding to an attack, the sire gradually is allowed to stay near the cat and can approach her from behind. Copulation with the isolated females only succeeded at the peak of oestrus, and only with a very patient sire; but immediately after copulation fierce agonistic behaviour was resumed.

We suggest that a harmonious equilibrium between the tendencies to attack and to flee is also important for the development of properly adjusted behaviour of the mother cat towards her young. The movements of the legs by which a mothercat catches her kitten or induces it to suckle and the movement by which it grips the kitten in the neck for transporting it, very much resemble similar, probably homologous, movements used in prey-catching and fighting (see also Leyhausen, 1956[a], p. 105). The important difference is that with respect to the young the movements are carried out in a very inhibited way. It seems likely to us that for becoming capable of grading the force appropriately through inhibition, experience like that obtained in "play" is necessary. The two kittens of the isolation experiment both killed and ate all kittens in their first litter. Although this also occurred in some cats that were not deprived during their development (we observed it in 3 of 10 first litters of mothers socially raised in the laboratory and never in many more births in the home situation) it seems more than co-incidence that one of the R-cats repeated devouring its litter 4 times, starting at the heads of the kittens as if they were prey. The other one had a second litter after it had been kept in the same cage with a very patient tomcat for a long period during which indications of fright and the frequency of agonistic behaviour dramatically declined. This cat took good care of her two kittens of the second litter and of three later litters as well.

Near their young kittens normal mothers are inhibited from killing prey which at a distance they will kill rapidly. An alarm vocalization by one of the kittens, however, may immediately remove the inhibition towards the prey.

Conflict hypothesis and ontogenetical development. The second possibility mentioned above about how social experience might influence the ontogeny of social behaviour suggests that the systems for aggression and escape, when activated, would directly influence the development of display patterns. This idea easily crosses ones mind when watching the agonistic play of littermates. One receives the impression that the various display postures are actually moulded during such encounters through interaction of the tendencies to attack and to excape.

Tinbergen's conflict hypothesis (Tinbergen, 1952; see for a recent critical discussion Baerends, 1975) assumes that a great many displays have in the course of evolution obtained their specific shape in this way. This hypothesis, therefore, was meant to give an ultimate explanation of the form of display. However, it is not unlikely that principles used in an ultimate explanation for the development of the displays of a species, or other taxonomic group, also apply to a proximate explanation for the development of displays during the ontogeny of the behaviour of each individual. Such a connection would not be surprising considering that phylogeny is actually modified ontogeny (de Beer, 1940). The fact that evolution works via the mechanisms of ontogeny does not necessarily imply, however, that in the course of evolution the schedule according to which the ontogenetical processes are programmed could not have changed. To provide for possible shortcuts introduced into the developmental proceedings as a display became more and more ritualized, the hypothesis of emancipation has been introduced. According to this idea a display originally derived from interaction between agonistic tendencies would gradually become detached from the systems originally motivating it and become incorporated into another behavioural system, for instance one subserving sexual or parental behaviour. In our opinion the evidence given so far for total emancipation of displays is extremely weak. On the contrary, the evidence for the conflict hypothesis in all cases investigated was derived from presently existing correlations between the occurrence of particular displays and overt expression of the attack and escape systems. A possibility – implying a compromise – is, however, that emancipation would take place in the course of the ontogeny of each individual as a result of learning processes (i.e. operant conditioning) after a primary moulding of the activity through the interaction of the tendencies to attack and to flee. As suggested earlier by Kruijt (1964) and by Feekes (1972), experience during ontogeny would give the performer of a particular posture the opportunity to associate it with the effect on its opponent. Reversely, the latter is at the same time in the position to associate this posture with accompanying overt expressions of attack and escape in its adversary or partner. Some more support for these ideas can be found in the literature. Sevenster (1968) has managed to teach stickleback males to use an "artificial" motor pattern (swimming through a ring or biting a rod) as an operant for obtaining a female. Through experiments with cats, who – in competition with conspecifics – had to run for food or solve a problem box situation, Winslow (1944[a,b]) has demonstrated the ability of cats to assess the predictability of responses of other, familiar, cats and take advantage of this knowledge. It seems worthwhile to us to consider some of our observations on the development of cat behaviour against the background of these ideas. In particular the occurrence of "rolling" is interesting. We have observed this motor pattern in the following different contexts:

(a) in a kitten as a result of being pushed over by the mother when she is licking it to remove feces or urine;

(b) in the kitten after the third week as an active invitation to the mother to approach and lick it;

(c) in a kitten during "play", apparently to invite the approach of another kitten;

(d) in a female in heat to invite the tomcat;

(e) in response to a passive or dead large prey (p. 75);

(f) in young and adults in extreme defence, in combination with kicking of the hindlegs.

Situation (a) occurs before the other ones (Pl. III[b]). This opens the possibility of explaining the origin of rolling in the situations (b), (c), (d) and possibly (e) as that of an operant in the Skinnerian sense. Then the reinforcing effect of the motherly care while the kitten was passively rolling over would have taught the kitten to use the rolling patterns actively for drawing attention, first with regard to the mother and later also generalized to other cats.

On the other hand the form of rolling can also be seen as an expression of simulateneously activated tendencies to flee, to attack and to stay on the spot, a combination of factors which in several cases has been satisfactorily used for explaining the shape of a display (Blurton Jones, 1968; Baerends, 1975). One observes that the dorsal part is turned away (escapes) from the opponent whereas the legs are attacking it (Pl. IV[e]). This hypothesis would nicely fit situation (f).

The conflict hypothesis could also serve as an ultimate explanation for the form of the rolling display in the situations (b), (c), (d) and (f). As a proximate explanation it would be less likely because we have no evidence that rolling occurs earlier in the life of the kitten in agonistic than in other situations. Moreover, the R-kittens, which were deprived of opportunities for agonistic behaviour, nevertheless when in heat performed rolling in response to tomcats. In situation (e) one may expect that agonistic tendencies and a tendency to attract were interacting. Much effort was needed to kill the relatively frightening prey, which approached the size of the littermates. In this situation rolling was especially shown by kittens who very much hesitated to attack the prey, but nevertheless stayed near to it (p. 64); a very mixed motivational condition is likely to have existed.

In our opinion it would be fruitful to entertain the idea that the form of the roling display has been derived in the course of phylogeny as the result of the incompatible tendencies to attack and to flee but that this display has to find its place in the behaviour of the individual kitten through the process of instrumental conditioning.

In the foregoing section we have advocated the opinion that the pattern of grasping and killing a prey becomes available to the kitten, some time after its maturation, after the perception of a definite feedback situation (reinforcement) as a consequence of opening up and eating the prey. It is imaginable that in a similar way an already matured display might become available. In that case the adequate feedback could be a special response of the opponent as well as a constellation of proprioceptive stimuli resulting from the assumption of a posture. Knowledge of the feedback to be expected must then be based on the genes but might be completed and/or modified by experience to a greater or lesser extent, possibly on the basis of an encoded programme. The reader is reminded here of the way the specific song pattern is produced and fixed in the white-crowned sparrow (Marler and Tamura, 1964; Konishi, 1965); viz. a programmed learning of the structure of the song pattern (template, "Sollwert") before the 4th month of life, followed by programmed efforts to produce a match of this template when the bird is about 8 months old and has become able to sing. After the bird has perceived the correspondance between the expectation and the performance of the vocal apparatus the song has become fixed.

Thus in the song bird, via this comparison, an otherwise endless, continuous developmental process is stopped and a particular developmental stage is conserved for the lifetime of the individual. Could it be that this procedure, with the many variations possible, would be more often used by the genes in transmitting species-specific features, including stereotyped social displays*? A similar idea has earlier been suggested by Wiley (1973) for the "strut" display of male sage grouse. A way to falsify this hypothesis is to look for cases in which the fixed pattern appears before the opportunity for acquiring it through exercise and shaping could have occurred. In the following we shall discuss information from the literature relevant in this context.

First, a number of cases have been reported of social or sexual motor patterns which could be induced in animals at a very early age, before the processes suggested here could have happened. Andrew (1966) induced precocious copulation (and tidbitting movements) in domestic chicks when a few days old by approaching them in a special way with some solid object. Schein and Hale (1959) obtained in male and female turkeys less than 2 months old copulation and strutting movements after treatment with androgen. According to Schleidt (1970) strutting is also present in young turkey chicks not treated in any special way.

* The problem of how in the individual a developmental process is stopped by proximate factors when a stage, which in the course of evolution has been determined by ultimate factors, is reached, equally exists for the morphogenesis of anatomical structures.

It is very unlikely that in the ontogeny of these activities experience with their effect can have played a role. They must have matured by mainly internal forces, such as we assumed for grasping and killing a prey, and were so to speak, waiting for the combination of internal and external factors capable of releasing them. Nevertheless, the evidence presented by these cases cannot be considered conclusive. The copulatory act is not a display. Strutting is used as a display in adult turkeys, but because it is also regularly performed in turkey chicks it is open to question whether it actually was precociously induced in the experiments of Schein and Hale; its occurrence might be functional at that time and this should be further studied. Andrew did not describe the precociously induced tidbitting in sufficient detail to convince that he really obtained the actual display, which is characterized by a specific call and specific orientations. The motor pattern of picking up objects is also used for feeding and other purposes.

A second sort of information could come from deprivation experiments but also this evidence is still scanty. Kruijt (1964) who has demonstrated in detail how side display and waltzing in jungle fowls is built up from elements of attack and escape (compromise activities; Andrew, 1956) nevertheless found them fully developed in cocks deprived during the first 6-12 months of their life from contact with conspecifics. If environmental factors would have promoted the development of these displays in the birds they could only have come from contacts with their caretakers or with inanimate objects. Pilot experiments in which the degree of isolation was increased even more by excluding direct contact with humans and by making the environment still more homogeneous, produced animals which were so extremely frightened when confronted with conspecifics that it was not feasible to test their potentiality for performing displays (Kruijt, pers.comm.).

A third type of evidence might come from the results of electrical stimulation of (subcortical) areas in the brain, which has in several cases produced display behaviour. For instance, Hunsperger, Brown and Rosvold (1964) and Hunsperger and Bucher (1967) report to have evoked forms of the arch and of growling and hissing in cats. Von Holst and Von Saint Paul (1960) obtained waltzing and crowing in domestic cocks and Delius (1973) could induce different types of threat behaviour from well defined loci in the brain of gulls. However, the fact that a behaviour pattern can be evoked this way does not exclude that it has developed through learning. By our knowledge, all subjects used in the experiments just mentioned had been together with conspecifics before. Delius even states that social experience in the period before the stimulation experiments started influenced their results: the electrodes most consistently evoking the most aggressive threat response were implanted in birds high in the social hierarchy, whereas electrodes implanted in gulls of low rank gave the most fearful threat responses. In recent years also for other subjects evidence has been obtained that experience influences the outcome of localized electrical brain stimulation. Moreover, the external situation during the test was found to be important for the type of activity obtained (see for a survey of the literature Kruk, Van der Poel and De Vos-Frerichs, 1979).

Consequently, at present this type of evidence cannot be used against the proposition that displays, the form of which has in the ultimate sense been determined by interaction of the tendencies to attack and to escape, are also in the proximate sense, i.e. in their ontogeny, developing under the influence of the antagonistic motivational systems concerned. We feel that at the present state of our knowledge one should stay open minded to the possibility that a display shaped in the course of evolution by the agonistic tendencies is in the ontogeny of each individual still developing under the influence of these factors (be it directly internal or via feedback from the environment) as well as to the alternative possibility that its ontogeny has become emancipated from the influence of attack and escape. In the latter case the problem of how the genes responsible for their stereotyped shape are coming to expression is not different from that in case of stereotyped "instrumental" activities, such as fixed action patterns subserving grooming, feeding, defecating, urinating, sleeping, copulation, etc.

The "stereotypy" of displays. Behaviour patterns are usually called stereotyped when they can easily be recognized in different individuals and sometimes also under different circumstances. This implies a certain amount of rigidity in the form of these patterns, but so far little research has been done on the problem how far actually this rigidity goes (Barlow, 1968; Schleidt, 1974).

Stereotyped displays have mainly been described for the lower vertebrates (birds, reptiles, fishes) and for invertebrates (arthropods, cephalopods). It has frequently been questioned even whether in mammals such "innate fixed action patterns" ("Instinkthandlungen", "Erbkoordination") acting as "social releasers", all in the sense of Lorenz (1935), actually exist. With regard to his study on communicative behaviour Schenkel (1948) has taken a shaded point of view. On the one hand he was able to distinguish (relatively small) stereotyped elements, on the other hand such elements always occurred in behavioural complexes with great communicative value, but highly variable in composition and appearance between different individuals and situations. Schenkel suggests an important influence of experience on the development of these complex communicative patterns. Because Weiss (1952) interpreted Schenkel's paper as a denial of the existence of fixed action patterns (which according to Weiss should by definition develop without influences from environmental information) in mammals, he undertook his social isolation experiments. He concludes that his results have proved the occurrence of fixed patterns in at least this mammalian species. However, it strikes that the patterns he found in his two semi-isolated cats (his experiments only excluded stimulation from conspecifics) mainly concern activities directly serving body maintenance, feeding, and copulation. The only communicative activities he lists are a defensive form of the arch and rolling.

In particular with respect to the arch we have stressed (p. 30) the versatility of its shape and orientation during encounters between kittens, but also during later life. This is in agreement with Leyhausen (1956[a]) who does not only explain the extreme arched-posture as the result of interaction of attack and escape (an interpretation comparable to that by Kruijt of the side display and waltzing of junglefowl), but also the entire gamut of postures and the transitions between them, shown in agonistic encounters of cats. Inspired by Lorenz' (1952) wellknown "roman square" with facial expressions of wolves Leyhausen pictures graphically the effect of the interactions of the aggressive and the fleeing tendencies on characteristics of the body contours and the face. There seems little doubt, but actually no hard proof, that these grades in posture and movements serve communicative effects. If so, we think it very unlikely that in such an effect no learning processes would be involved; our data on this, however, does not go beyond the abnormalities in the behaviour of the two semi-isolated R-kittens when meeting another cat (p. 83).

From this discussion it seems to emerge that two types of display activities might be distinguished: (a) stable stereotyped ones, and (b) more flexible forms, which are transitional between two stereotyped extremes. In mammals the (b)-type seems to be of considerable importance and is likely to increase in significance the more flexible the body must be for ecological reasons (carnivores), and in general the higher the mammal is developed (facial expressions in human beings). The (a)-type has always been emphasized for invertebrates and lower vertebrates, particularly for birds. This type of posture certainly jumps to the eye, is much easier to describe, and gives a foothold for phylogenetic considerations (Lorenz, 1935). For the black-headed gull Tinbergen (1959) has given some evidence that transitional forms between different (a)-type displays are relatively rare. Recent work (Baerends, Van Rhijn and Veen, unpubl.) on the communicative displays of the little and the black-headed gull, and also some remarks made by Delius (1973) regarding the result of his brain stimulation experiments in gulls, give the impression that the frequency of transitional forms may have been underestimated and possibly also their importance for communication.

Release, orientation and context of communicative activities. The third category of effects we expected of deprivation experiments with kittens (see p. 82) were influences on the release and/or the orientation of activities. Release concerns the situations in which different activities are applied and the objects towards which the activities are directed. Both were found to be influenced by deprivation and so was the effectiveness of the orientation with which the activities were carried out. In the earlier mentioned deprivation experiments of Harlow on rhesus monkey and Kruijt on jungle fowl the occurrence (= release, or recognition of object or situation) and orientation of social responses were also found to have been affected.

The situation in which a display is performed, its orientation and also its temporal relations in sequence with other activities of communicative value may have a considerable influence on the type of reaction of the receiver. It is likely that the knowledge the latter has of the significance of a particular display in different contexts develops by means of learning processes. The rolling display, already discussed on p. 85, is an example of such a dipslay which in different contexts serves different messages meant for different receivers.

IV.5. IMITATION

As mentioned in the introduction, for the origin of species-specific fixed patterns, like those occurring in mouse-catching, innateness and imitation have often been considered as the only alternatives. In this study arguments were found for a role of learning processes in the development of these activities. In the following we shall briefly discuss whether or not a concept of imitation would fit to any of these learning processes.

It should first be realized that no unanimity exists on the definition of imitation. Warden and Jackson (1939) define as "intelligent imitation" the immediate performance of the identical act after watching the behaviour of another animal without practice. On the other hand Thorndike (1911) has defined imitation as the formation of an association between the sight of an object and the act towards it, through an idea gained by watching. An important difference between these definitions is that whereas the latter one includes cases in which an animal, after having picked up the idea by watching a conspecific, achieves successful performance itself by the incorporation of trial and error learning; the former excludes this possibility. Another important difference between definitions handled by different authors is in the exactness of the imitation required for a motor pattern to be called imitative. In this Thorndike, for instance, is much more strict than Berry (1908).

Definitions of different authors also vary on the extent to which an association between the act performed by another animal and the reward it obtained has been made by the "imitating" animal. If such an association is likely Berry speaks of "voluntary" imitation. In contrast he postulates two kinds of "involuntary" imitation: "instinctive" imitation and "automatic" (or "habitual") imitation. The latter would both be caused by a tendency to join a conspecific in performing the same behaviour, either an innate pattern or an acquired habit respectively. The terms "social facilitation" (Klopfer, 1959, 1961) and "contagious behaviour" (Thorpe, 1963) have also been used in this sense. Berry is of the opinion that "involuntary imitation" (thus e.g. social facilitation and local enhancement) is instrumental in acquiring an effective behaviour pattern more or less resembling the example observed. Sometimes imitation does not involve the behaviour pattern performed but only the localisation; the animal exploring in the area where another one is or has been busy, but in its own trial and error way (local enhancement, Klopfer, 1959). For the entire gamut of possibilities mentioned here the terms social learning and observational learning have been introduced.

Miller and Dollard (1941) in an attempt to make the concept of imitation objectively operational, have distinguished between matched-dependent behaviour in which cues from a leader determine which behaviour pattern an individual performs at a particular moment, and actual copying in which the individual responds to cues of sameness and difference produced by stimulation from his own responses and those of the model.

This short survey of the tremendous spread in the interpretation of the term "imitation" will suffice here as a background against which the various statements in the literature about imitation in cats have to be regarded. For a recent more thorough and critical review of the imitation concept see Davis (1973).

Most experiments on observational learning in cats are concerned with solving problems presented by various kinds of apparatus. In several cases (Thorndike, 1911, Berry, 1908; Herbert and Harsh, 1943; John, Chesler, Bartlett and Victor, 1968) watching another cat solving the problem certainly had a facilitating effect on the later performance of the observers. In most cases, however, the example was not exactly copied. This was a reason for Thorndike to deny the occurrence of imitation in cats, but not for Berry who did not take correct copying as a criterion for imitation and speaks of "voluntary imitation of a lower order" (p. 20). John et al. demonstrated the possibility of transfer of information about the kind of behaviour patterns to be carried out in two series of experiments. In these cats were given an opportunity to observe already trained cats in a situation in which they avoided an electric shock by jumping a hurdle at the appearance of an auditory stimulus, and in another situation in which they could obtain food by pressing a lever when a particular visual stimulus was shown. When later placed in these experimental situations the observer cats learned the tasks significantly quicker than fully naive controls; some of the "observers" even performed correctly at the first opportunity. It cannot be said, however, neither for the hurdle jump nor for the lever pressing, that the behaviour pattern executed was shaped after the model observed. From the available repertoire a behaviour pattern matching the one observed was chosen. In our opinion the results obtained by John et al. (1968) would fit better the matched-dependent than the copying category.

Still, in the results of Berry and of Herbert and Harsh correctly copied behaviour elements do seem to have occurred. It is interesting that the latter authors moreover found that watching not only the perfect execution of a learned habit but also the proceedings during learning in another animal, including its incorrect manipulations, was more beneficial to acquiring the demonstrated pattern than merely watching the end result.

Kittens are certainly interested in what their littermates and their mother are doing. They often concurrently show the same behaviour pattern and are also often simultaneously attracted to the same source of stimulation (Pl. V^{e}). In this way an opportunity is created for the transmission of information about the recognition and handling of prey animals. Berry's observations (1908) of a mother cat with three kittens, to which he offered live mice, indicate that social facilitation is involved.

Kuo (1930) has compared the incidence of rat killing between kittens kept from the 3rd week on in isolation to whom between the 2nd and the 16th week a rat was presented every fourth day, and kittens kept as a litter with their mother, who saw their mother in a separate area catch a rat – but not eat it – before they were themselves presented with a rat (in separation from the mother) on a similar schedule as the first group. Both groups consisted of 20 kittens. Of the group which had seen the example 17 kittens became rat killers, of the other group 9. Unfortunately the groups not only differ with regard to the presence or absence of the opportunity to watch the example. The kittens who had this possibility also had social experience and this we have found to influence prey catching positively (p. 58). In another experiment meant to investigate the possibility of imitation, Kuo (1938) gave 16 kittens, who after having been kept with adult white rats for 4½ months had remained indifferent to them, the possibility of seeing a rat-killing cat kill and eat a white rat in the same cage. The kitten tested got the opportunity to kill a rat itself before and after the example was given. The tests were repeated every day for two weeks. At the end of this period 6 kittens had attempted to kill the white rat which was their former cage mate, 3 did so with success. To us the results of both experiments taken together indicate that, by means of an example, information can be passed about the potentiality of an object to serve as a prey. However, the experiments give no evidence about the transmission of information on how to perform the necessary behaviour patterns.

We ourselves did the following experiment with regard to a possible occurrence of observational learning in prey-catching.

Two kittens trained to catch mice in the early period were from the age of 118 days confronted once a week for 10 min. with the aquarium containing water and a swimming fish. At day 175, after 10 presentations, they had failed to catch a fish. They were interested and followed the movements of the fish attentively. But after they had a number of times wetted their nose and paws they gave up every attempt to reach the fish from the water surface and watched it through the glass.

From then on the kittens were joined by an experienced fish-catcher after having been in the arena with the aquarium fish for 5 minutes. They had free opportunity to watch the latter at work, catching and eating the fish. They never interfered but after the fish-catcher had lost interest they sniffed at the spot where it had eaten the fish, eagerly licking at some remains. When eating was finished the "teacher" was removed and another fish was put into the aquarium; then the "pupil" received another 10 min. opportunity.

The experiment was concluded on day 350. One kitten who had watched fish catching 22 times in no case angled a fish from the water. When fish were given on dry ground, however, it rapidly ate them. The second kitten, after having observed 10 performances by the experienced cat, angled its first fish on day 272. Until day 350, 18 more fish were presented to it, of which 15 were angled. However, this angling was fully different from the example; the mouse-catcher seized the fish with both forelegs while standing on its hindlegs outside the aquarium and bending over it. The kitten was able to do it this way because of its relatively large size and strength at the age it started catching fish.

Relatively correct copying of behaviour patterns seems to occur more frequently in primates. The examples of cultural transmission of newly invented manipulations for cleaning food in Japanese monkeys have especially become famous (Itani, 1958; Kawai, 1965; Tsumori, 1967). However, the most correct copying of a behaviour pattern has been found in the ontogeny of bird song (see p. 86). One wonders whether memorizing a tune is not in general an essential aid in the different task of memorizing a motor pattern and copying it (dance!).

One of the few cases in which copying of visually demonstrated motor patterns could be established also concerns a bird species. Dawson and Foss (1965) showed that budgerigars are able to copy the way in which they have seen another budgerigar remove a lid from a food trough.

We do not find it surprising that imitation in the sense of copying is so rare. How would the imitator know which of all manipulations it has been watching is really essential for achieving the goal and which not? It has to make its own instruction or to receive instruction given intentionally, preferably also verbally, as is so far only known in humans. However, also in humans actual detailed copying of behaviour is a difficult task as every instructor knows, and as for instance has been described in detail by Miller and Dollard (1941) for the case of a poorly singing adult who ardently wished to learn to carry a tune.

IV.6. PLAY

Notwithstanding all difficulties to define play (see e.g. Loizos, 1966, 1967; Aldis, 1975; Bekoff, 1976), the playing of kittens is proverbial. Although in the above we have in general avoided the term "play" we did from time to time yield to the temptation to use it as a collequialism. Activities such as attempts to catch butterflies, touching and throwing objects (or even prey), wrestling, and chasing will be casually labelled as play by all non-sophisticated observers. It is, therefore, worthwhile to ask whether our view derived from the study of the ontogeny of cat behaviour might help to clarify the nature and eventually also the concept of play.

Attempts to clarify the concept should concern the functional and the causal aspects of play. So far more attention has been paid to the former than to the latter aspect (see Bekoff, 1976). A possible function acknowledged by most authors on the subject is the acquisition of information and of skills through experience resulting from the practice obtained during playing. In several instances we have been able to find an effect of practising, e.g. when comparing the effect of training on mouse-catching with that of training on fish-catching, the absence of stalking in kittens kept in the laboratory, the behaviour differences between kittens reared in isolation and reared as a litter, etc. Consequently we conclude that practice, and thus also play – because it provides opportunities for practice – is an effective instrument of behavioural ontogenesis.

If one accepts this conclusion one might expect that in the course of time play activities would gradually merge into "serious behaviour" with a direct or immediate function. It is probably with this idea in mind that some studies were undertaken of the course of the frequency of playing activities during the developmental period of kittens. Egan (1972) did so for object play. She found an increase for the time spent playing in the first 8-12 weeks followed by a decrease which started earlier the lower the effectiveness of the object for eliciting play. For the "best" object (the small real fur roll) the decline began after 24 weeks. One may conclude from this that the threshold for object play is lowest in the 8-12th week and then rises gradually. Egan found the temporal change in frequency to be the same for all play activities she had distinguished. West (1974) has determined similarly the course of the frequency of social play activity with the age of the kittens. She found a peak at 12 weeks and a decline from the 16th week on. An optimum curve for the course of play-frequency with age was also found by Koepke and Pribram (1971) in experiments primarily designed to study possible differences in sucking behaviour of kittens sucking either from a normal lactating cat or from an anesthetized nonlactating one. In the milk-sucking group the play-peak lies between the weeks 12 and 18; in the dry-sucking group between the weeks 8 and 12, playing developing at a higher rate. This difference is probably causally related with the difference in decline of sucking between both groups, which was considerably steeper in the dry-sucking kittens. Unfortunately the authors do not mention which activities were labelled as play! The beginning of play is in all studies assumed to occur around the 4th week. We are of the opinion that it is fully arbitrary where one assumes "play" to start in the continuous developmental process; it depends on which activities one chooses to define as play.

A comparable study was also made by Barrett and Bateson (1978) on the frequencies of 7 well defined activities: cat contact, object contact, arch, neck-flex, wrestle, stalk, and rear between the 4th and 12 th week. Cat contact, arch, neck flex, and rear were found to reach their highest frequencies around the 6th and 7th week; object contact, stalk, and wrestle went up in frequency until the 12th week, the end of the observation period. This shift with time of the relative frequency of these different activities cannot be interpreted as the result of the early-peak activities developing into the ones with a late frequency peak. It is far more likely that cat contact, arch, neckflex, and rear are frequent in the early period because then the kittens of a litter tend to stay near each other. In the later period they have become more ambulant. This makes more stalking possible and also more object play. Wrestling tends to occur at the end of a run and this fits West's data that chasing is more common at 12 then at 6 weeks of age. The difference in mobility is also likely to account for the absence found of correlations between the frequencies of most play measures in the period from 4-7 and from 8-12 weeks.

West found that the 8 activities she distinguished in play, viz. pounce, side-step, the complementary positions of wrestling opponents: belly up and stand-up, face off, vertical stance (= rear), horizontal leap and chase, tend to take up different positions in sequences of social play; statistically they tend to follow each other in the order in which they are given here. She also reports that the number of such sequences increased from 2.1 per kitten per 45 min. at 5 weeks to 9.4 sequences in the period between 8 and 16 weeks.

Summarizing the evidence from these quantitative data, they indicate that play activities increase in frequency with the gaining of strength of the kittens from the 4th up to the 8th week, that from the 8th to the 16th week the threshold for play is relatively low, and that this threshold gradually increases from around the 16th to the 32th week. During this period the formation of sequences of play activities and the form of the flexible displays (b-displays, see p. 89) seem to be developing. Our own observations have shown an improvement in the orientation of the activities, an increase of the efficiency of their use, and a gradual stabilizing of the interactions between attack and escape.

Since in most activities performed during play a typical stereotyped form is easily recognizable (cf. Loizos, 1966, 1967), since at some time in the life of the animals these activities serve in an adapted way in one or more functional contexts, and since (with the exception of b-displays) they tend to keep this basic form with age (although they may ''grow''; Würdinger, 1970), many play activities may be classified as fixed action patterns*. When Lorenz (1937) introduced this concept, using the terms ''Erbkoordination'' and ''Instinkthandlung'' for it, his most important arguments for distinguishing this type of activity separately were the phylogenetically determined stereotyped form of the activities and the independance of this form of external stimulation during performance. In addition Lorenz has postulated as a third essential criterium that the elicitability of each of these activities by external stimuli would depend on the strength of a specific (variable) internal factor of its own.

Following this latter idea, Leyhausen (1965) assumes the existence of such an internal factor for each fixed action pattern subserving predatorial behaviour (e.g. lying in wait, stalking, chasing, seizing, killing and eating). This would give these activities a fair amount of independence and each of them might serve as appetitive behaviour with regard to the other (relative hierarchy of moods). Leyhausen strongly rejects the idea of a superimposed unitary ''prey-catching drive''. His main argument is the one which had earlier inspired Lorenz to formulate his ''action-specific energy'' concept, namely that in satiated cats a number of predatorial action patterns can still be released. In our opinion, however, this matter can only be properly discussed on the basis of a systematic investigation of the behavioural organization of preycatching in its entirety, which – at least to our knowledge – does not yet exist. Such an organization has to be adaptive and may therefore vary from species to species. For instance, in connection with a study on surplus killing in carnivores, Kruuk (1972[b]) has argued that it is not unlikely that the observations referred to by Lorenz and Leyhausen can be interpreted as the result of special adaptations to enable a species capable of storing food to acquire a surplus when food is easily available, as a reserve for less favourable times.

It has been stated as well as questioned that in play the activities would occur

* When using this term we do not intend any implications about the developmental processes through which in the individual the characteristic form of a fixed action pattern is reached. We are also aware that variations may occur between different performances in the same individual as well as between performances of different individuals (Schleidt, 1974), a reason for Barlow (1968) to propose the term modal action pattern.

in a more arbitrary order than in directly functional behaviour and that they would more often be repeated (see Loizos, 1966; Bekoff, 1976). To our knowledge Eibl-Eibesfeldt (1951) was the first to suggest (for agonistic play in squirrels) on such ground that in play the fixed action patterns might operate independently of a superordinated system. This idea has also been expressed by Leyhausen (1965) with regard to the play of kittens and in connection with his concept of relative hierarchy, and by Barrett and Bateson (1978) who say that "the lack of association between many of the categories suggest that output from a variety of separately controlled systems may be commonly lumped under the general heading of play" (p. 116).

Our impression from watching kittens play is that during play periods external stimuli are more dominating in determining at each moment which activity occurs than during periods of serious behaviour, in which different patterns usually occur temporally linked in groups controlled by common causal factors (see Baerends, 1976). It is, however, extremely difficult to systematically collect reliable quantitative data to test this subjective impression, mainly because of the problem of monitoring the totality of releasing stimuli perceptible for the kittens at each moment. So far our impression is mainly based on the frequent occurrence of switches in behaviour which do not seem directly functional and which can nearly always be correlated with the appearance of a for us conspicuous external stimulus. If the occurrence of the various activities during play would really be mainly determined by external stimuli this implies that during that period the influence on these activities of superimposed mechanisms would be relaxed. Egan's (1972) finding that the frequency of object play was not influenced by mild food-deprivation seems to argue in favour of this idea. This can be the case when all the primary needs of the animal are satisfied and when the escape behaviour is not activated by fear stimuli from outside the home area of the kittens. The motivational mechanism bringing about the relaxation of ties with systems of higher order one might call the "play-system". This system, therefore, would not be in control of a specific set of activities but exert an influence on other systems as a result of which these systems temporarily lose that control. There are indications that special signals may control its release and maintenance in conspecifics. In primates the "play face" is considered to be such a signal (Loizos, 1967); West (1974) has suggested that particular tail movements may serve such a function in cats; Bekoff (1974) mentions several examples in other mammals.

When the play-state is released in an animal, the internal state of the disinhibited subsystems and the presence of releasing stimuli specific for each of these systems will determine which one at a moment will be expressed. In chapter III and IV we have repeatedly stated that young kittens more often put their full force into an activity than older ones. It is probably because of this that Loizos mentions "exaggeration"as a characteristic of play; the impression that play activities often look exaggerated results from the fact that adult animals hardly perform their activities without inhibition. We have shown how through the interaction between attack and escape – or approach and withdrawal (but usually more elements are involved than could simply be accounted for by these opposite orientation components) – the kittens learn to assess the amount of force that has to be put into an activity. We also found that the escape factor (e.g. a big prey or a slightly disturbing external situation) postponed the performance of the full pattern and gave the kitten an opportunity for more exploratory and manipulatory behaviour, i.e. it facilitated the occurrence of play. Consequently we agree with Aldis (1975) when he states that play is characterized by an interaction between the tendencies to attack and to escape. It is also the influence of the inhibiting escape tendency which so often makes a play activity incomplete, another character of play mentioned by Loizos.

We assume that on the basis of the possibility for the fixed action patterns to be released in dissociation of superimposed control mechanisms during play as well as the inhibited character of their performance, the opportunity is created for conditioning processes leading to improvements of orientation and of discrimination between various sources of stimuli. As a result the activities will be more effectively performed. Moreover the animal will learn to apply them in an effective order. The essence of this idea was brought forward by Kortlandt (1955) in connection with the ontogeny of what he calls the "hierarchical system of appetites", on the basis of his observations on the behaviour of young cormorants (Kortlandt, 1940). Following the ideas of Kruijt (1964), and with Harlow's (1969) findings in Rhesus monkeys on the effects of playing with peers on later social and sexual behaviour in mind, we further suggest that the experience during play "teaches" the kittens how to deal with the interacting influences of the opposed tendencies to attack and to flee, an important prerequisite for socialization.

With regard to the causation of play we feel no need to distinguish between object play and social play; we rather think in terms of one play inducing mechanism serving the development of behaviour with different functions.

It has been mentioned by Loizos, West, and others that play is self-rewarding i.e. a positive feedback leads to continuation or repetition. It can and need not be stimulated by the conventional type of rewards, it is a reward in itself. Loizos (1967) makes the remark that "to regard play as practice for adult function does not account for the fact that adults as well as infants play in most mammalian species in which play occurs at all" (p. 185). She also says that she could think of other and perhaps more efficient mechanisms for obtaining practice than play. However, remarks like this latter one can be made for any morphological structure, behaviour pattern, and physiological or behavioural mechanism. Their nature cannot be understood on the basis of the present demands in the organism under investigation alone, one has also to take the possibilities and restrictions of the phylogenetical history into account.

As Loizos has said, play is a relatively recent "invention" in evolution. It seems to have especially developed in groups such as carnivores and primates, where flexibility in behavioural sequences and the acquisition of experience have a considerable survival value, in particular for obtaining food, However, when play develops under the influence of one selection factor, such as food, its function will easily extend to other functional contexts, such as social behaviour. We suggest that the play mechanism developed with the increase of the role of learning in the ontogeny of behavioural control systems superimposed on the fixed action patterns. Disinhibition of the elementary patterns through activation of the play system provides the possibility for exercise and learning at a time when the knowledge and skill to be gained is not yet needed, and during a period when play activity is relatively safe, both because the young are still under parental protection.

The play state is obviously mainly controlled by factors active in youth, but this need not exclude that it can also be released in adults, for instance by strong external stimulation (highly preferred play objects or vivid play behaviour of the own young). In cases where it would be of survival value it could easily be imagined that a mechanism for releasing the play state in adults could be built in. The relative independence of hunting behaviour from eating in carnivores mentioned above, might be due to such a mechanism; when the supply of prey is irregular it probably pays for the predator not to miss a chance. In the higher primates there seems to exist a tendency to exploit the mechanism permitting the disinhibition of fixed action patterns to an increasing extent for exploration of the environment, culminating in the capacity of experimenting with one's biotope and remodelling it in man.

V. Summary

CHAPTER I

Building upon the existing literature a study was undertaken to investigate the ontogeny of the social and predatorial behaviour of the domestic cat. The question is asked how far the ontogeny of this behaviour is influenced by programmed learning during development. An attempt is made to attack the problems by carefully following the changes in the morphology of the behaviour of developing kittens and by experimentally manipulating experience, in particular with regard to prey-catching.

CHAPTER II

The behaviour of kittens in different stages of their development and the way it gradually changes with age are described in connection with corresponding changes in the complex of environmental stimuli impinging upon the kittens. Such stimuli are considered as potential candidate-factors for giving shape to the behaviour of the developing young cats. In the first month they result from interactions with the mother, with littermates, and with the nest. In the course of time they are increasingly provided by animate and inanimate stimuli, at first from the immediate surroundings of the nest and later from a larger area. Three behaviour patterns of the mother: nosing and licking, treading and scraping, and other, more narrowly aimed movements with the legs and mouth are thought to be in particular important for stimulating and guiding the young in the early stage, and for directing them towards that area of her body where the nipples are situated.

The early development of locomotory behaviour goes through three phases. In the first phase the body musculature dominates, enabling the kitten to make oscillating movements, the legs following passively. In the second phase the body is still dominating but the legs become capable of supporting it, whereas at the end of the third phase the legs can move fairly independently of the activity of the body musculature.

In the course of these phases dorsoventral curving of the spine increases in frequency and power. A 4-week old kitten is physically able to catch and kill a small adult mouse. At that time all sensory organs are fully functional. Whereas during the first two weeks stimulation of the kitten mostly evoked approach, in kittens over that age retreat was more often elicited, in particular by rapidly approaching visual stimuli. From then on characteristics of approach or attack and of withdrawal or escape can be distinguished in the activities performed by interacting littermates. In the course of further development the inhibitive force of these tendencies appears to increasingly cause a graded performance of approach and attack. It is suggested that displays such as the arch and the vertical stance or rear seem to result from this interaction; in locomotion it leads to stalking and in the manipulation of objects in the environment to an adequate assessment of the timing of activities and the amount of force exerted. Also the orientation of the movements is gradually improved. During this second month individual differences in the performance of behaviour sequences become apparent. Correlations of such differences with differences in experience make it likely that in their causation learning is involved. Relatively weak inhibitive forces were found to induce "play" and in that way create opportunities for learning.

CHAPTER III

After a brief description of the prey-catching behaviour in adult kittens a more detailed description is given of the gradually developing behaviour towards prey of the growing kitten and of the ways this behaviour could be influenced by provisioning activities of the mother. Consecutively, the problem how far experience of the kittens during this developmental phase might affect their adult predatorial behaviour, was experimentally investigated.

In two series of experiments kittens were allowed to obtain experience with only one species of prey between the 4th and the 8th week. The kittens, alone or with littermates, with the mother present or absent, were offered daily live mice in Series I A and swimming fish in Series IB, both in standard situations. Their behaviour towards the prey and its progressive changes with time were watched and recorded. One purpose of these series was to obtain insight into possible opportunities for acquiring knowledge about prey; another was to study the influence of the littermates and the mother on the development of the catching and killing patterns.

Only 3 of the 40 kittens in Series I did not achieve killing a mouse within 56 days. The earliest killing took place at the 27th day, most kittens first killed a mouse when between 35 and 47 days. The probability that a kitten kills a mouse increased considerably after the killing of the first mouse. For 53% of the kittens after their first killing the killing ratio (number of mice killed/number of mice presented) exceeded 0.70 in the experimental situation.

The presence of littermates has a positive influence on the killing ratio because it tends to reduce the tendency to escape; however, interactions between littermates over the prey are rare. Under the experimental conditions the presence of the mother tended to postpone the date of first killing, mainly because she interferes and kills a prey caught by a kitten when the latter is slow in finishing with it. However, the behaviour under these circumstances varies between individual mothers at least partly as a result of their own experience as kittens.

Kittens are capable of catching and killing mice without any previous exerience with this prey. However, observations in a larger experimental space strongly suggest that the expression of this capacity by the kittens is normally facilitated by the reassuring effect of a mother being quietly present and sometimes by her interference, particularly in the case of big or relatively ferocious prey. This situation helps the kittens to adjust their catching and killing patterns to the particular circumstances, for instance with regard to timing, force exerted, and tactics.

Six out of seven kittens confronted with fish swimming in a tray managed in the course of the tests of Series I B to angle the fish from the water with one leg. The main problem for the kittens in these tests was to overcome the aversion of getting wet and in particular to wet their noses. The earliest fish catching accomplished in these tests took place at the age of 49 days; after the first success no failures were observed. Just as in Series I A with mice all kittens with fish as prey also finally arrived at similar stereotyped patterns for catching, killing and eating the prey. Differences in experience were expressed in the appetitive behaviour leading to the release of these patterns.

In six different experiments of Series II the behaviour towards mice and fish of the kittens trained in Series I in a later period (56-280 days) was compared with the predatorial behaviour of the kittens of other litters which during the early period (4-8 weeks, see Table 6) had only obtained amorphous food, and of one litter which had access to a large garden during its first 14 weeks. From the beginning of the tests in Series II the kittens were kept without the mother; in one experiment the litters consisted of only one kitten kept in isolation, in the others the kittens were kept together in litters of 3 or 4.

Kittens which had killed mice in the early period and which had been raised as litters were very successful mouse-catchers in the later period, independently of whether they had been trained in the presence or in absence of the mother. The achievement of the unexperienced laboratory-raised kittens varied considerable. This must be attributed to considerable individual differences in the time elapsing before the first mouse is caught and killed; as soon as this had happened all kittens became efficient mouse-catchers. In all kittens with experience in the garden situation it took a relatively large number of tests before the first mouse was caught, killed and eaten under laboratory conditions (after which they too proved themselves to have become efficient mouse-killers). These kittens had acquired the habits of approaching a prey carefully and transporting it to a safe place, as a consequence of which they performed less well within the 10 min. permitted them during each test. Kittens trained on fish were at once quick in catching mice; killing and eating, however, took place less frequently than in kittens experienced with mice. This was probably due to these kittens not being used to a prey capable of escaping after it had been caught. Only one of 12 kittens trained to catch mice ever became a fish-catcher. This low success is due to the problems of how to cope with the water barrier.

Neither of the two kittens reared in isolation managed to kill and eat a mouse; only once a mouse was caught, but it was not retained. These kittens showed fright responses with regard to the prey.

The conclusion is drawn that the motor patterns for seizing and killing a mouse develop (mature) independently of experience with prey. To apply them adequately, however, experience with prey is important, experience with littermates also seems to serve this purpose. Experience has a function in organizing the appetitive behaviour of catching and in confirmation of the catching and killing acts as soon as the kitten has been rewarded by eating.

Before catching, between catching and killing, between killing and opening up the prey, and while eating "interruptive" activities may be performed which do not seem to directly subserve feeding purposes. Their incidence increases when a kitten for some reason is inhibited in continuing its predatory behaviour. Grooming is an interruptive behaviour pattern particularly occurring before catching. Carrying, touching, and tossing chiefly occur after the prey has been caught. They are likely to contribute to the acquisition of skill in handling the prey and have often been classified as "play". Head rubbing and rolling with respect to prey could be seen as an attempt to attract it, similar to the way in which play mates and sexual partners are invited by means of these behaviour patterns.

CHAPTER IV

In this chapter an attempt is made to combine the observations and data given in the chapters II and III with relevant data from the literature into a general frame of how the morphogenesis of the behaviour of the domestic cat in its totality may be organized.

Arguments are given that the development of the core of all behaviour patterns, the locomotory sigmoid body movement with the alternating leg movements accompanying it, proceeds through maturation, i.e. is mainly internally controlled. Experience is thought to become increasingly important when after the 4th week the movements of the extremities have attained a considerable independance from those of the body musuculature. Although in a kitten of one month the fixed action patterns of catching and killing and those of the appetitive behaviour leading to these acts have sufficiently matured to be potentially operational, the first successful killing and eating is usually postponed as a result of fright releasing stimuli emanating from the prey. This inhibition of attack, which at the same time develops in confrontations with littermates leads to the frequent performance of behaviour of the fixed action pattern type in a form which is generally considered to be characteristic for play. It is suggested that this behaviour is additionally facilitated by a mechanism bringing about a relaxation of the ties of the fixed action patterns with control-systems of higher order. The execution of these activities during play opens possibilities for obtaining experience with stimuli and objects in the environment as well as with the potentialities of the own body and the effects of the own behaviour on the environment. As a result of these experiences the individual builds up individually preferred behaviour sequences, which can serve predatory, social, or other functions. With regard to predatory behaviour such sequences tend to become established as soon as a prey has been killed and eaten for the first time. Some characteristics of potential food (rapid movement, creaking sound, fish odour) seem to release predatorial behaviour without previous experience. With respect to other characteristics conditioning is likely to take place, for instance when the food has been brought by the mother to the kittens in or near the nest.

In "playful" interactions between littermates a gamut of body postures is displayed which can be understood as the result of interaction of the tendencies to attack and to flee. It is suggested that play might serve to give the kittens an opportunity to learn the communicative value (message and meaning) of displays, particularly of graded displays. Further, the possibility is discussed that even displays with a phylogenetically fixed stereotyped form might in the ontogeny of the individual be brought about via programmed learning during play.

Two kittens reared in absence of other cats from the 7th week on – and thus deprived of possibilities for social play – later showed signs of insufficient harmonious control of their attack and escape tendencies in agonistic, sexual, and parental encounters.

The possible role of "imitation" (observational learning) in the morphogenesis of motor patterns in kittens is discussed. Copying the behaviour of another kitten was never observed, only indications for "local enhancement" or "social facilitation" were obtained.

REFERENCES

Adamson, J. – Born free. Collins and Harvill, London. 1960.

Adler, H. E. – Some factors of observational learning in cats. J. Genet. Psychol. **86**, 159-177 (1955).

Aldis, O. – Play fighting. Academic Press, New York. 1975.

Andrew, R. J. – Some remarks on behaviour in conflict situations, with special reference to Emberiza sp. Brit. J. Anim. Beh. **4**, 41-45 (1956).

Andrew, R. J. – Precocious adult behaviour in the young chick. Anim. Behav. **14**, 485-500 (1966).

Baerends, G. P. – A model of the functional organization of incubation behaviour. In: The Herring Gull and its Egg. G. P. Baerends and R. H. Drent (Eds.), Behaviour Suppl. **17**, 263-312 (1970).

Baerends, G.P. – An evaluation of the conflict hypothesis as an explanatory principle for the evolution of displays. In: Essays on Function and Evolution of Behaviour. G. P. Baerends, C. Beer and A. Manning (Eds.), 187-227 (1975). Clarendon Press, Oxford.

Baerends, G. P. – The functional organization of behaviour. Anim. Behav. **24**, 726-738 (1976).

Baldaccini, N. E. – An ethological study of reproductive behaviour, including the colour patterns of the Cichlid fish Tilapia mariae (Boulanger). Monitore zool. ital. (N.S.) **7**, 247-290 (1973).

Barlow, G. W. – Ethological units of behaviour. In: Central nervous systems and fish behavior. D. Ingle (Ed.) 217-232 (1968). University of Chicago Press.

Barrett, P. and P. Bateson – The development of play in cats. Behaviour 66, 106-120 (1978).

Beer, G. R. de – Embryos and ancestors. Clarendon Press, Oxford. 1940.

Bekoff, M. – Social play and play-soliciting by infant canids. Amer. Zool. **14**, 323-340 (1974).

Bekoff, M. – Animal play: problems and perspectives. In: Perspectives in Ethology. P. P. G. Bateson and P. H. Klopfer (Eds.). Vol. 2, 165-188 (1976). Plenum, New York.

Berry, C. S. – An experimental study of imitation in cats. J. of Comp. Neurol. Physiol **18**, 1-12 (1908).
Blurton Jones, N. G. – Observations and experiments on causation of threat displays of the Great Tit (Parus major). Anim. Beh. Monogr. **1**, **2**, 75-158 (1968).
Carmichael, L. – The development of behaviour in vertebrates experimentally removed from the influence of external stimulation. Psychol. Rev. **33**, 51-58 (1926).
Carmichael, L. – A further study of the development of behavior in vertebrates experimentally removed from the influence of external stimulation. Psychol. Rev. **34**, 34-47 (1927).
Coghill, G. E. – Anatomy and the problem of behaviour. Cambridge Univ. Press. 1929.
Cooper, J. B. – A description of parturition in the domestic cat. J. Comp. Psychol. **37**, 71-79 (1944).
Davis, J. M. – Imitation: a review and critique. In: Perspectives in Ethology. P. P. G. Bateson and P. H. Klopfer (Eds.). (1973). Plenum Press, New York, London.
Dawson, B. V. and B. M. Foss – Observational learning in budgerigars. Anim. Behav. **13**, 470-474 (1965).
Delius, J. – Agonistic behaviour of juvenile gulls, a neuroethological study. Anim. Behav. **21**, 236-246 (1973).
Eaton, R. L. – The predatory sequence, with emphasis on killing behavior and its ontogeny, in the Cheetah (Acinonyx jubatus Schieber). Z.f. Tierpsychol. **27**, 492-504 (1970).
Eaton, R. L. – An experimental study of predatory and feeding behavior in the Cheetah (Acinonyx jubatus). Z.f. Tierpsychol. **31**, 270-280 (1972[a]).
Eaton, R. L. – Predatory and feeding behavior in adult lions: the deprivation experiment. Z.f. Tierpsychol. **31**, 461-473 (1972[b]).
Egan, P. J. – Factors influencing object-play in young domestic cats. Ph. D. Dissertation. University of Cambridge. 1972.
Egan, P. J. – Object-play in cats. In: Play. J. S. Brunner, A. Jolly and K. Sylva (Eds.). (1926). Basic Books, New York.
Eibl-Eibesfeldt, I. – Beobachtungen zur Fortpflanzungsbiologie und Jugendentwicklung des Eichhörnchens. Z.f. Tierpsychol. **8**, 370-400 (1951).
Eibl-Eibesfeldt, I. – Zur Biologie der Iltis (Putoris putorus L.). Zool. Anz. Suppl. **19**, 304-314 (1956).
Eibl-Eibesfeldt, I. – Angeborenes und Erworbenes im Verhalten einiger Säuger. Z. Tierspsychol. **30**, 705-754 (1963).
Eisenberg, J. F. and P. Leyhausen – The phylogenesis of predatory behavior in mammals. Z. Tierpsychol. **30**, 59-93 (1972).
Elliot, J. P., I. McTaggart Cowan and C. S. Holling – Prey capture by the African Lion. Can. J. Zool. **55 (11)**, 1811-1828 (1977).
Ewer, R. F. – Suckling behaviour in kittens. Behaviour **15**, 146-162 (1960).
Ewer, R. F. – Further observations on suckling behaviour in kittens together with some general considerations of the interrelations of innate and acquired responses. Behaviour **17**, 247-261 (1961).
Feekes, F. – "Irrelevant" ground pecking in agonistic situations in Burmese red junglefowl (Gallus g. spadiceus). Behaviour **43**, 186-326 (1972).
Golani, I. – Homeostatic motor processes in mammalian interactions in a choreography of display. In: Perspectives of Ethology 2. P. P. G. Bateson and P. H. Klopfer (Eds.), 69-134 (1976). Plenum Press, New York, London.
Gray, J. – How animals move. University Press, Cambridge. 1953.
Harlow, H. F. – Age-mate or peer affectional system. In: Advances in the Study of Behavior. D. S. Lehrman, R. A. Hinde and E. Shaw (Eds.). Vol. **2**, 333-383. (1969). Acad. Press, London, New York.
Harlow, H. F. and M. K. Harlow – Social deprivation in monkeys. Sci. Amer. **207**, 137-146 (1962).
Harlow, H. F., M. K. Harlow and E. W. Hansen – The maternal affectional system of rhesus monkeys. In: Maternal behavior in mammals. H. L. Rheingold (Ed.) 254-281 (1963). Wiley, New York.

Herbert, M. J. and Ch. M. Harsh – Observational learning by cats. J. Comp. Psychol. **37**, 81-95 (1944).

Hogan, J. A. – An experimental study of conflict and fear: an analysis of behavior of young chicks toward a mealworm. Part I: The behaviour of chicks which do not eat the mealworm. Behaviour **25**, 45-97 (1965).

Hogan, J.A. – Idem II. The behaviour of chicks which eat the mealworm. Behaviour **27**,273-289 (1966).

Holst, E. von – Die relative Koordination als Phänomen und als Methode Zentralnervöser Funktionsanalyse. Erg. Physiol. **42**, 228-306 (1939).

Holst, E. von and U. von Saint Paul – Vom Wirkungsgefüge der Triebe. Naturwiss. **18**, 409-422 (1960).

Hunsperger, R. W., J. L. Brown and H. E. Rosvold – Combined stimulation in areas governing threat and flight behaviour in the brain stem of the cat. Progr. in Brain Res. **6**, 191-197 (1964).

Hunsperger, R. W. and V. M. Bucher – Affective behaviour produced by electrical stimulation in the forebrain and brain stem of the cat. Progr. in Brain Res. **27**, 103-127 (1967).

Itani, J. – On the acquisition and propagation of a new food habit in the natural group of the Japanese Monkey at Takasaki-Yama. Primates **1**, 84-98 (1958).

John, E.R., Ph. Chesler, F. Bartlett and I. Victor – Observation learning in cats. Science **159** 3822), 1489-1491 (1968).

Kawai, M. – Newly-acquired pre-cultural behaviour of the natural troop of Japanese monkeys on Koshima Islet. Primates **6**, 1-30 (1965).

Klopfer, P. H. – Social interactions in discrimination learning with special reference to feeding behaviour in birds. Behaviour **14**, 282-299 (1959).

Klopfer, P. H. – Observational learning in birds. Behaviour **17**, 71-80 (1961).

Koepke, J. E. and K. H. Pribram – Effect of milk on the maintenance of sucking behavior in kittens from birth to six months. J. Comp. Physiol. Psychol. **75**, 363-377 (1971).

Konishi, M. – The role of auditory feedback in the control of vocalization in the white-crowned sparrow. Z.f. Tierpsychol. **22**, 770-783 (1965).

Kortlandt, A. – Eine Uebersicht der angeborenen Verhaltensweisen des Mitteleuropäischen Kormorans (Phalacrocorax carbo sinensis Shaw & Nodd) ihre Funktion, ontogenetische Entwicklung und phylogenetische Herkunft. Arch. Néerl. Zool. **4**, 401-442 (1940).

Kortlandt, A. – Aspects and prospects of the concept of instinct. Arch. Néerl.Zool. **11**, 155-284 (1955).

Kruijt, J. P. – Ontogeny of social behaviour in Burmese Red Junglefowl (Gallus g. spadiceus). Behaviour Suppl. **12**, 1-201 (1964).

Kruijt, J. P. – Early experience and the development of social behaviour in Junglefowl. Psychatr., Neurol. Neurochir. **74**, 7-20 (1971).

Kruk, M. R., A. M. van der Poel and T. P. de Vos-Frerichs – The induction of aggressive behaviour by electrical stimulation in the hypothalamus of male rats. Behaviour, **68**, (1979).

Kruuk, H. – The spotted Hyena. Chicago Univ. Press. 1972[a].

Kruuk, H. – Surplus killing by carnivores. J. Zool. Lond. **166**, 233-244 (1972[b]).

Kruuk, H. and M. Turner – Comparative notes on predation by lion, leopard, cheetah and wild dog in the Serengeti area, East Africa. Mammalia **31**, 1-27 (1967).

Kuo, Z. Y. – The genesis of the cat's response to the rat. J. Comp. Psychol. **11**, 1-35 (1930).

Kuo, Z. Y. – Further study on the behavior of the cat towards the rat. J.Comp. Psychol. **25**, 1-8 (1938).

Lehrman, D. S. – A critique of Konrad Lorenz's theory of instinctive behavior. Quart. Rev. Biol. **28**, 337-363 (1953).

Leyhausen, P. – Ueber die Beziehung der Katze zum Beutetier. Verh. d. deutschen Zool. Ges Freiburg, 200-202 (1952)

Leyhausen, P. – Verhaltensstudien an Katzen. Z. Tierpsychol. Beih. **2**, (1956[a], 1973).

Leyhausen, P. – Das Verhalten der Katzen (Felidae). Hb. d. Zool. **7**, **10**, 1-34 (1956[b]). de Gruyter, Berlin.

Leyhausen, P. – Ueber die Funktion der relativen Stimmungshierarchie (dargestellt am Beispiel der phylogenetischen und ontogenetischen Entwicklung des Beutefangs von Raubtieren). Z. Tierpsychol. **22**, 412-494 (1965).

Leyhausen, P. und R. Wolff – Das Revier einer Hauskatze. Z.f. Tierpsychol.**16**, 666-670 (1959).

Loizos, C. – Play in mammals. In: Play, exploration and territory in mammals. P. A. Jewel and C. Loizos (Eds.). Symp. Zool. Soc. London **18**, 1-9 (1966).

Loizos, C. – Play behaviour in higher primates: a review. In: Primate ethology. D. Morris (Ed.) 176-218 (1967). Weidenfeld and Nicolson, London.

Lorenz, K. – Der Kumpan in der Umwelt des Vogels. J. Ornithol. **83**, 137-213, 289-413 (1935).

Lorenz, K. – Ueber die Bildung des Instinktbegriffes. Naturwiss. **25**, 289-300, 307-318, 324-331 (1937).

Lorenz, K. – Die Entwicklung der vergleichenden Verhaltensforschung in den letzten 12 Jahren. Verh. d. Deutsch, Zool. Ges. Freiburg, 36-58 (1952).

Lorenz, K. – Phylogenetische Anpassung und adaptive Modifikation des Verhaltens. Z. Tierpsychol. **18**, 139-187 (1961).

Lorenz, K. – Evolution and modification of behavior. Univ. Press, Chicago, London, 1-121. 1965.

Lorenz, K. and N. Tinbergen – Taxis und Instinkthandlung in der Eirollbewegung der Graugans. Z. Tierpsychol. **2**, 1-29 (1938).

Marler, P. and M. Tamura – Culturally transmitted patterns of vocal behaviour in Sparrows. Science **146**, 1483-1486 (1964).

Miller, N.E. & J. Dollard – Social learning and imitation. New Haven, Yale University Press (1941).

Moynihan, M. – Some aspects of reproductive behavior in the Black-headed Gull (Larus ridibundus ridibundus L.) and related species. Behaviour Suppl. **4**, (1955).

Nottebohm, F. – Ontogeny of bird song. Science **167**, 950-956 (1970).

Prechtl, H. F. R. and W. M. Schleidt – Auslösende und steuernde Mechanismen des Saugaktes I. Z. vgl. Physiol. **32**, 257-262 (1950).

Prechtl, H. F. R. and W. M. Schleidt – Auslösende und steuernde Mechanismen des Saugaktes II. Z. vgl. Physiol. **33**, 53-63 (1951).

Prechtl, H. F. R. – Angeborene Bewegungsweisen junger Katzen. Experientia **8**, 220 (1952).

Preyer, W. – Spezielle Physiologie des Embryos; Untersuchungen über die Lebenserscheinungen von der Geburt. Grieben, Leipzig. 1885.

Rasa, O. A. E. – Prey capture, feeding techniques, and their ontogeny in the African Dwarf Mongoose, Helogale undulata rufula. Z. f. Tierpsychol. **32**, 449-488 (1973)

Rosenblatt, J. S. and T. C. Schneirla – The behaviour of cats. In: The behaviour of domestic animals. E. S. E. Hafez (Ed.) (1962). Baillière, Tindall and Cox, London.

Rosenblatt, J. S., G. Turkewitz and T. C. Schneirla – Early socialization in the domestic cat as based on feeding and other relationships between female and young. In: Determinants of infant behaviour. B. M. Foss (Ed.) 51-74 (1959) London.

Rosenblatt, J. S., G. Turkewitz and T. C. Schneirla – Development of home orientation in newly born kittens. Trans. New York Acad. of Sciences Ser. II, **31**, **3**, 231-250 (1969).

Schaller, G. B. – The deer and the tiger. Univ. Press, Chicago. 1967.

Schaller, G. B. – The Serengeti Lion. Univ. Press, Chicago. 1972.

Schein, M. W. and E. B. Hale – The effect of early social experience on male sexual behaviour in androgen injected turkeys. Anim. Behav. **7**, 189-200 (1959).

Schleidt, W. M. – Precocial sexual behaviour in turkeys (Meleagris gallopavo L.). Anim. Behav. **18**, 760-761 (1970).

Schleidt, W. M. – How "fixed" is the fixed action pattern. Z. Tierpsychol. **36**, 184-211 (1974).

Schenkel, R. – Ausdruck-Studien an Wölfen. Behaviour **1**, 81-130 (1948).

Schneirla, T. C. – An evolutionary and developmental theory of biphasic processes underlying approach and withdrawal. In: Nebraska Symposium on Motivation. 1-42 (1959). Univ. of Nebraska Press, Lincoln.

Schneirla, T. C., J. S. Rosenblatt and E. Tobach – Maternal behavior in the cat. In: Maternal Behavior in Mammals. L. Rheingold (Ed.) (1963). Wiley & Sons, New York.

Sevenster, P. – Motivation and learning in sticklebacks. In: Central nervous systems and fish behaviour. D. Ingle (Ed.), 233-245 (1968). University of Chicago Press.

Thomas, E. & F. Schaller – Das Spiel der optisch isolierten, jungen Kaspar-Hauser-Katze. Naturw. **23**, 557-558 (1954).

Thorpe, W. H. – Learning and instinct in animals. 2nd. ed. Methuen, London. 1963.

Thorndike, E. L. – Animal intelligence. MacMillan, New York. 1911.

Tinbergen, N. – Derived activities: their causation, biological significance, origin and emancipation during evolution. Quart. Rev. Biol. **27**, 1-32 (1952).

Tinbergen, N. – Comparative studies of the behaviour of Gulls (Laridae): A progress report. Behaviour **15**, 1-70 (1959).

Tsumori, A. – Newly acquired behavior and social interaction. In: Social Communication among Primates. S. A. Altmann (Ed.) 207-219 (1967). Univ. Press, Chicago.

Uexküll, J. von – Umwelt und Innenwelt der Tiere. Springer, Berlin. 1909.

Weiss, G. – Beobachtungen an zwei isoliert aufgezogenen Hauskatzen. Z. Tierpsychol.**9**, 451-462 (1952).

West, M. – Social play in the domestic cat. Amer. Zool. **14**, 427-436 (1974).

Warden, G. J. and T. A. Jackson – Imitative behavior in the rhesus monkey. J. Comp. Psychol. **27**, 133-147 (1939).

Wiley, R. Haven – The strut display of male sage grouse: a "fixed action pattern" Behaviour **47**, 129-152 (1973.)

Winslow, Ch. N. – The social behavior of cats. I. Competitive and aggressive behavior in an experimental runway situation. J. Comp. Psychol. **37**, 297-313. (1944[a]).

Winslow, Ch. N. – Idem II. Competitive, aggressive and food sharing behavior when both competitors have access to the goal. J. Comp. Psychol. **37**, 315-326. (1944b).

Würdinger, I. – Erzeugung, Ontogenie und Funktion der Lautäusserungen bei vier Gänse-arten (Anser indicus, A. caerulescens, A. albifrons und Branta canadensis). Z. f. Tierpsychol. **27**, 257-302 (1970).

Yerkes, R. M. and D. Bloomfield – Do kittens instinctively kill mice? Psychol. Bull. **7**, (8), 253-263 (1910).

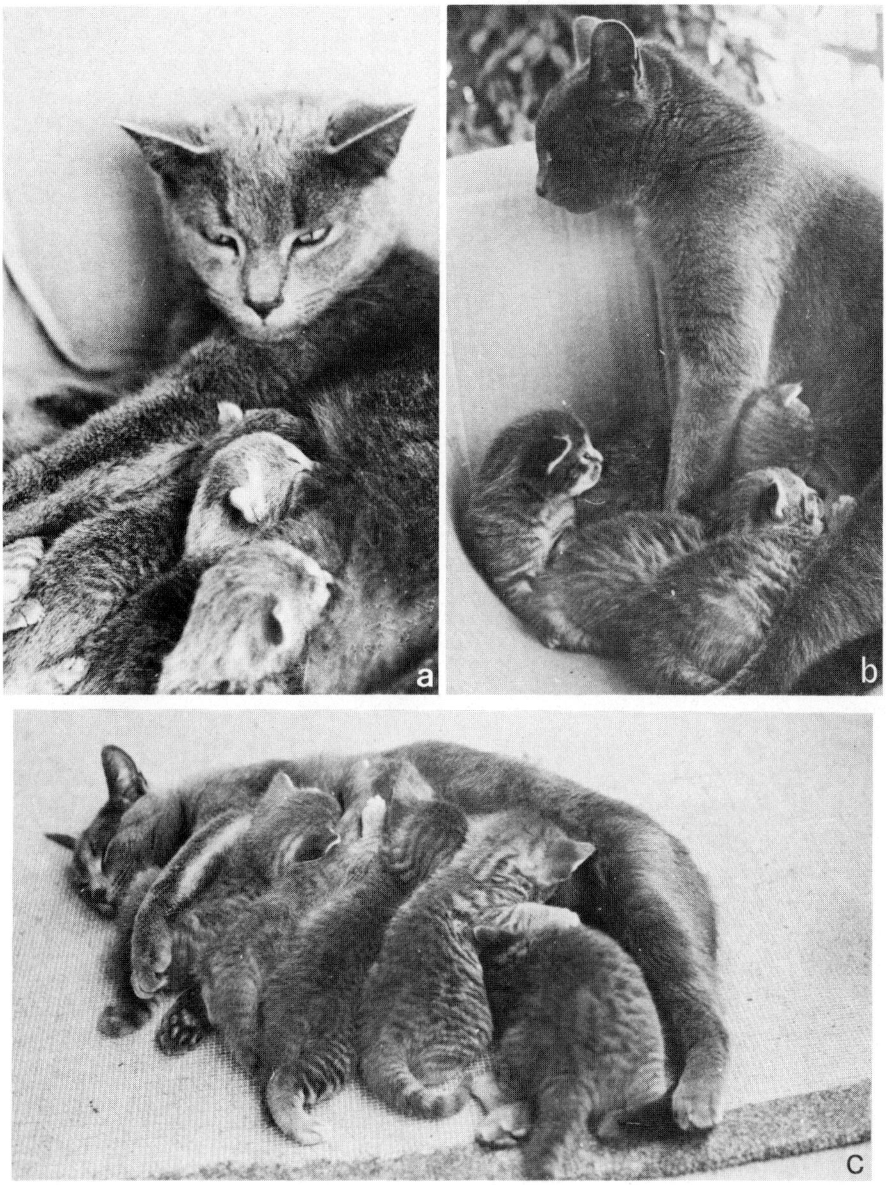

PLATE I.

Pl. I. Feeding mother a) 2nd day; b) 15th day; c) 28th day.

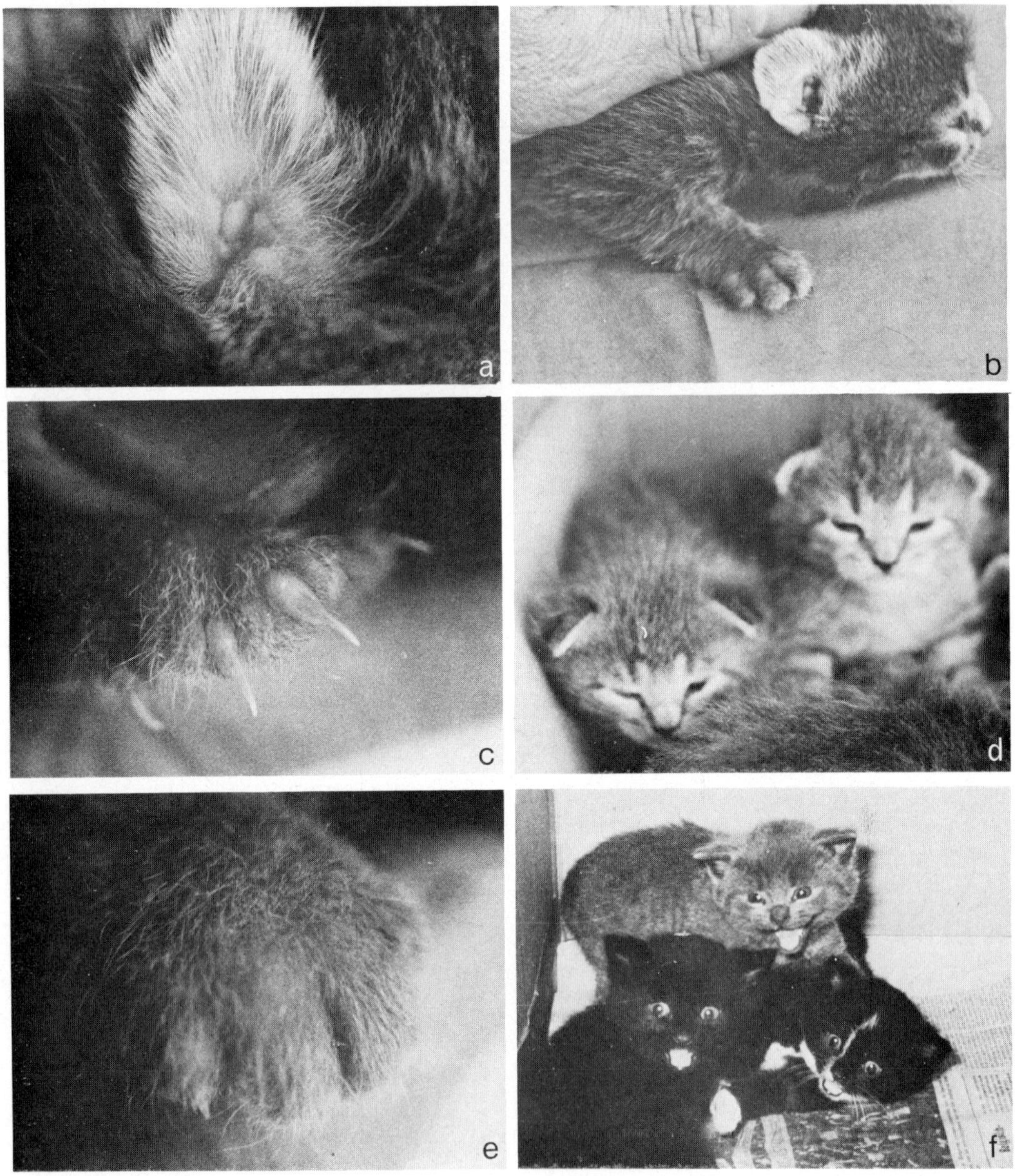

PLATE II.

Pl. II. Early developments. a) pinna with auditorial canal closed, 2nd day; b) with open canal, 10th day; c) forefoot with nails extended and, e) with nails contracted, 3rd week; d) 15-day old kittens; f) 3-weeks old kittens, hissing.

PLATE III.

Pl. III. Upgrowing young. a) kitten on 2nd day; b) mother licking anal region of 8-day old kitten; c) kitten, 23rd day; d) mother with litter, 15th day; e) kitten, 12th week; f) mother with 10-week old kittens.

PLATE IV.

Pl. IV. Agonistic play behaviour (8th week). a) and b) arch during approach and passing; c) mutual attacking, followed by (d) appease through laying down; e) fight, left kitten in defense by rolling over; f) attack followed by escape.

PLATE V.

Pl. V. Social interactions (8th week). a) agonistic meeting between kittens; b) friendly meeting between mother and young; c) threatening approach (note thick tail); d) fleeing; e) stalking together; f) pouncing on littermate.

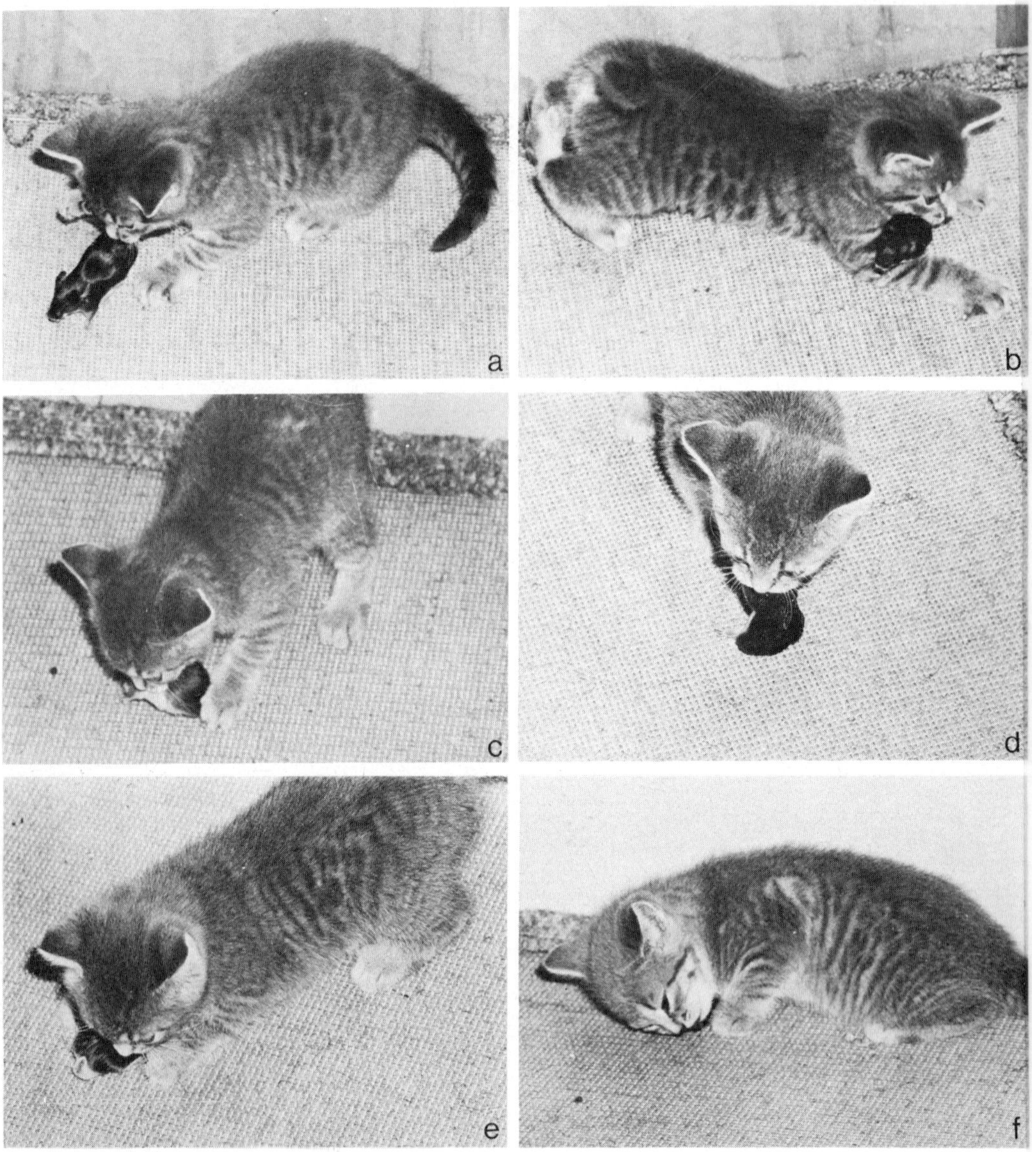

PLATE VI.

Pl. VI. Successful mouse catcher (6th week); note extended nails. a) following; b) attack fails; c) catching; d) killing; e) eating; f) snipping.

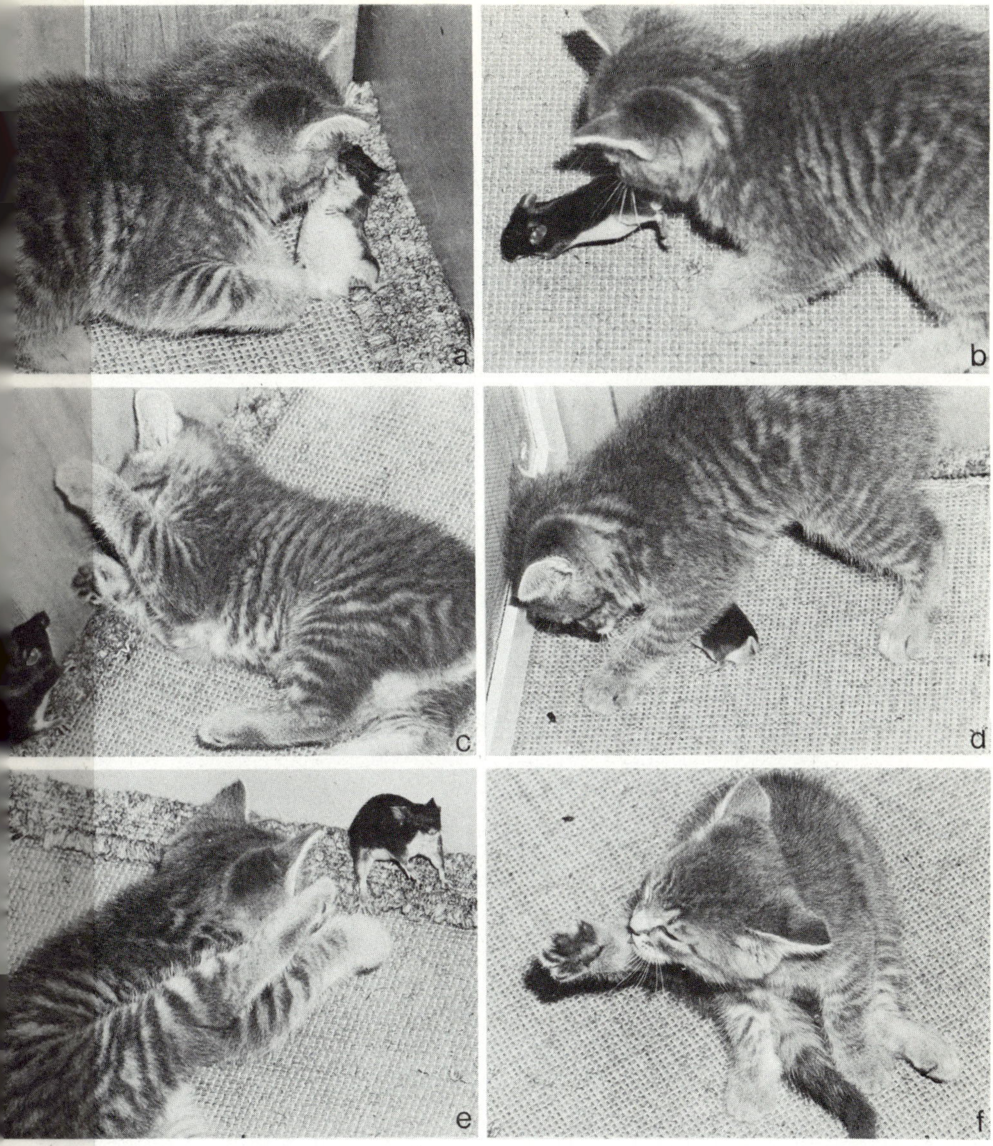

PLATE VII.

Pl. VII. Unsuccessful mouse catcher (6th week); note protracted nails. a) mouse threatens; b) awkward catch; c) rolling and slamming; d) stepping over mouse; e) begin of rolling after grooming; f) grooming and ignoring mouse.

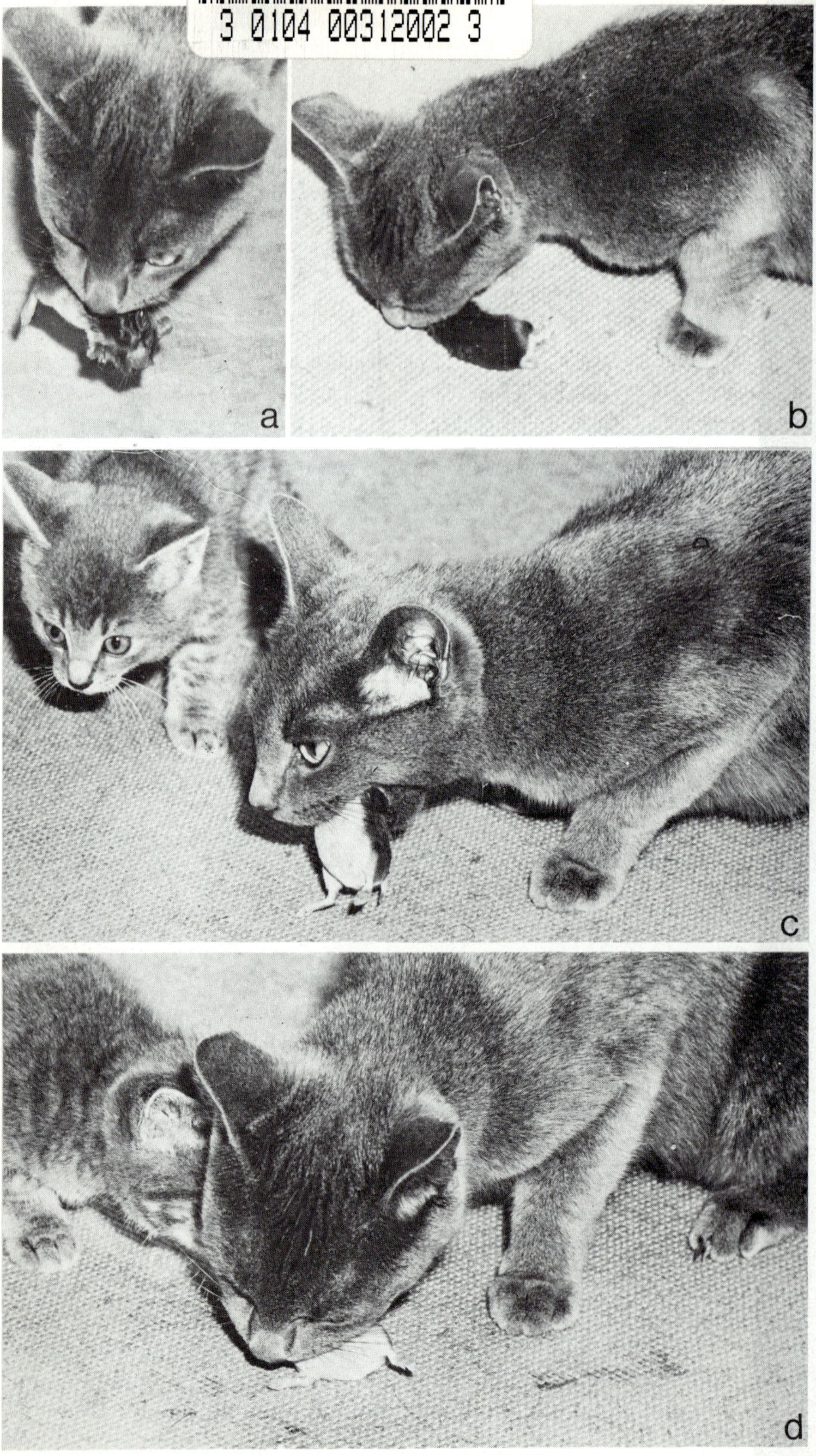

PLATE VIII.

Pl. VIII. Provisioning a 6-week old young. Mother (a) catches, (b) kills, and (c) presents mouse to young; d) young joins eating.